MEASUREMENT UNCERTAINTY AND PROBABILITY

A measurement result is incomplete without a statement of its 'uncertainty' or 'margin of error'. But what does that statement actually tell us? This book employs the principles of measurement, probability and statistics to describe what is meant by a '95% interval of measurement uncertainty' and to show how such an interval can be calculated.

The book argues that the concept of a 'target value' is essential if the principles of probability are relevant. It advocates the use of 'extended classical' statistical methods, such as (i) the propagation of higher moments of error distributions, (ii) the evaluation of 'average confidence' intervals, and (iii) the evaluation of 'conditional confidence' intervals. It also describes the use of the Monte Carlo principle for simulating measurements and for constructing procedures that give valid uncertainty statements.

Useful for researchers and graduate students, this book promotes the correct understanding of the classical statistical viewpoint. It also discusses other philosophies, and it employs clear notation and language to avoid the confusion that exists in this controversial field of applied science.

ROBIN WILLINK is a physicist and mathematical statistician who has been employed in Applied Mathematics at Industrial Research Ltd, the parent body of the National Metrology Institute of New Zealand. He is the author of many articles on the subject of measurement uncertainty, and he has made several notable contributions to the statistics literature.

MEASUREMENT UNCERTAINTY AND PROBABILITY

ROBIN WILLINK

CAMBRIDGE UNIVERSITY PRESS
Cambridge, New York, Melbourne, Madrid, Cape Town,
Singapore, São Paulo, Delhi, Mexico City

Cambridge University Press
The Edinburgh Building, Cambridge CB2 8RU, UK

Published in the United States of America by Cambridge University Press, New York

www.cambridge.org
Information on this title: www.cambridge.org/9781107021938

© R. Willink 2013

This publication is in copyright. Subject to statutory exception
and to the provisions of relevant collective licensing agreements,
no reproduction of any part may take place without the written
permission of Cambridge University Press.

First published 2013

Printed and bound in the United Kingdom by the MPG Books Group

A catalogue record for this publication is available from the British Library

Library of Congress Cataloguing in Publication data
Willink, Robin, 1961–
Measurement uncertainty and probability / Robin Willink.
pages cm
Includes bibliographical references and index.
ISBN 978-1-107-02193-8
1. Measurement uncertainty (Statistics) 2. Probabilities. I. Title.
QA276.8.W56 2012
519.2–dc23
2012025873

ISBN 978-1-107-02193-8 Hardback

Cambridge University Press has no responsibility for the persistence or
accuracy of URLs for external or third-party internet websites referred to
in this publication, and does not guarantee that any content on such
websites is, or will remain, accurate or appropriate.

To my friend Gemma, who hasn't got a clue what this book is about.
But that doesn't matter to either of us.

Contents

Acknowledgements		*page* xi
Introduction		xii

Part I Principles

1	Foundational ideas in measurement	3
	1.1 What is measurement?	3
	1.2 True values and target values	5
	1.3 Error and uncertainty	7
	1.4 Identifying the measurand	10
	1.5 The measurand equation	12
	1.6 Success – and the meaning of a '95% uncertainty interval'	16
	1.7 The goal of a measurement procedure	19
2	Components of error or uncertainty	21
	2.1 A limitation of classical statistics	21
	2.2 Sources of error in measurement	23
	2.3 Categorization by time-scale and by information	28
	2.4 Remarks	30
3	Foundational ideas in probability and statistics	32
	3.1 Probability and sureness	33
	3.2 Notation and terminology	39
	3.3 Statistical models and probability models	43
	3.4 Inference and confidence	45
	3.5 Two central limit theorems	52
	3.6 The Monte Carlo method and process simulation	53
4	The randomization of systematic errors	56

4.1	The Working Group of 1980	57
4.2	From classical repetition to practical success rate	59
4.3	But what success rate? Whose uncertainty?	63
4.4	Parent distributions for systematic errors	65
4.5	Chapter summary	70

5 Beyond the ordinary confidence interval — 72
- 5.1 Practical statistics – the idea of average confidence — 72
- 5.2 Conditional confidence intervals — 77
- 5.3 Chapter summary — 80

Part II Evaluation of uncertainty

6 Final preparation — 85
- 6.1 Restatement of principles — 85
- 6.2 Two important results — 87
- 6.3 Writing the measurement model — 88
- 6.4 Treating a normal error with unknown underlying variance — 92

7 Evaluation using the linear approximation — 98
- 7.1 Linear approximation to the measurand equation — 98
- 7.2 Evaluation assuming approximate normality — 103
- 7.3 Evaluation using higher moments — 105
- 7.4 Monte Carlo evaluation of the error distribution — 120

8 Evaluation without the linear approximation — 125
- 8.1 Including higher-order terms — 125
- 8.2 The influence of fixed estimates — 130
- 8.3 Monte Carlo simulation of the measurement — 135
- 8.4 Monte Carlo simulation – an improved procedure — 141
- 8.5 Monte Carlo simulation – variant types of interval — 145
- 8.6 Summarizing comments — 150

9 Uncertainty information fit for purpose — 152
- 9.1 Information for error propagation — 152
- 9.2 An alternative propagation equation — 157
- 9.3 Separate variances – measurement of functionals — 161
- 9.4 Worst-case errors – measurement in product testing — 164
- 9.5 Time-scales for errors — 169

Part III Related topics

10 Measurement of vectors and functions — 175
- 10.1 Confidence regions — 176

	10.2	Simultaneous confidence intervals	179
	10.3	Data snooping and cherry picking	184
	10.4	Curve fitting and function estimation	185
	10.5	Calibration	189
11	Why take part in a measurement comparison?		192
	11.1	Examining the uncertainty statement	192
	11.2	Estimation of the Type B error	198
	11.3	Experimental bias	201
	11.4	Complicating factors	203
12	Other philosophies		204
	12.1	Worst-case errors	204
	12.2	Fiducial inference	205
	12.3	Bayesian inference	207
	12.4	Measurement without target values	214
13	An assessment of objective Bayesian statistics		220
	13.1	Ignorance and coherence	222
	13.2	Information and entropy	230
	13.3	Meaning and communication	234
	13.4	A presupposition?	235
	13.5	Discussion	236
14	*Guide to the Expression of Uncertainty in Measurement*		237
	14.1	Report and recommendation	237
	14.2	The mixing of philosophies	239
	14.3	Consistency with other fields of science	241
	14.4	Righting the *Guide*	242
15	Measurement near a limit – an insoluble problem?		245
	15.1	Formulation	245
	15.2	The Feldman–Cousins solution	247
	15.3	The objective Bayesian solution	250
	15.4	Purpose and realism	253
	15.5	Doubts about uncertainty	255
	15.6	Conclusion	257
Appendix A	The weak law of large numbers		259
Appendix B	The Sleeping Beauty paradox		260
Appendix C	The sum of normal and uniform variates		261
Appendix D	Analysis with one Type A and one Type B error		262
Appendix E	Conservatism of treatment of Type A errors		263

Appendix F	An alternative to a symmetric beta distribution	264
Appendix G	Dimensions of the ellipsoidal confidence region	266
Appendix H	Derivation of the Feldman–Cousins interval	267
References		268
Index		273

Various quotations at the beginnings of Parts I, II and III were obtained from "Statistically speaking: a dictionary of quotations", selected and arranged by C. C. Gaither and A. E. Cavazos-Gaither. The last is due to Paul Gardner, writer-producer of "Richard Tuttle: Never Not an Artist".

The author can be contacted at robin.willink@gmail.com

Acknowledgements

This book is borne out of a respect for the ideas of truth and meaning. That respect is associated with my trust in God, whom I see as having given me the motivation, ability and courage required. The material contained has been informed by many past and present discussions with scientists at the Measurement Standards Laboratory of New Zealand; I am especially grateful for the support of Blair Hall, Jeremy Lovell-Smith, Rod White and Annette Koo. I would also like to thank an anonymous referee of my proposal to Cambridge University Press about this book. In submitting a report, the referee wrote to the publisher 'You offered me to choose a Cambridge book: if it is allowed to me, I will choose exactly a copy of this book!' What a wonderful, affirming, remark.

Introduction

I have recently finished a period of employment as a statistician at a research institute that, in one capacity, acts as the parent body for New Zealand's national laboratory for metrology (measurement science). During that period, a significant amount of time was spent considering and discussing ideas of data-analysis in physics and chemistry, especially in matters relating to measurement. I quickly reached the conclusion that the ideas surrounding the quantification of measurement uncertainty are problematic, and later I formed some opinions as to why that is the case. First, the evaluation of uncertainty is supposed to have been adequately addressed in the *Guide to the Expression of Uncertainty in Measurement* (BIPM et al., 1995; JCGM, 2008a) but – as is described in Chapter 14 – that document is unsatisfactory from a statistical point of view. Second, for typical measurement scientists, the evaluation of uncertainty in measurement is not the most interesting aspect of their work. Less effort is expended in understanding conceptual matters of uncertainty than in addressing practical matters of measurement technique. Third, and perhaps most important, the task involves concepts of statistical inference that might not have been presented well during an education in a science faculty.

The statistician involved in this area of science might reasonably expect to find one outstanding difficulty: the 'Bayesian controversy' that splits the statistical community over the nature of 'probability'. This controversy is surely relevant in measurement problems, where systematic errors are ubiquitous and where these errors might be treated by imputing worst-case values – as a classical statistician often would – or be treated 'probabilistically' – as a Bayesian statistician would. But the statistician might be surprised to also find another difficulty: the measurement community seems divided over whether quantities can be considered to have 'true' values. That issue is of great relevance. The ground is cut out from underneath someone carrying out parameter estimation if a 'true value' for the parameter does not exist. Yet the concepts of such parameter estimation are those on which measurement scientists appear to base their treatments of measurement uncertainty. Thus, there is this curiosity: some measurement scientists are implicitly denying the very basis of the methods of data analysis that they employ.

So the statistician seeking to make sense of what metrologists understand by 'the quantification of measurement uncertainty' has a surprisingly difficult task. This was emphasized to me when, in the library at my place of work, I viewed a copy of the book *Measurement Errors: Theory and Practice* by the experienced and authoritative Russian scientist Semyon Rabinovich (1995). At three places in the margin of the very first page, where Rabonivich introduces basic ideas of measurement, some dissenting reader had written the word 'wrong'! No further evidence was needed to convince me that different measurement scientists think about measurement in fundamentally different ways. If such scientists cannot agree on foundational ideas about measurement then how can the statistician contribute easily in the area of 'measurement uncertainty'? Perhaps some of the uncertainty in measurement is about the concept of measurement itself!

Despite these difficulties – or perhaps because of them – my work at that institute led to the publication of a number of articles of a statistical nature in international measurement journals. The kind words and compliments that I received in response to those articles indicated that many measurement scientists are searching for a coherent understanding of measurement uncertainty based on accepted logical principles. A time came when a colleague in Slovakia asked me 'Robin, have you thought of writing a book on this subject?' I had, but I had not made or taken an opportunity. The opportunity has since presented itself, and this book is the result.

The purpose of the book

I am a physicist-turned-statistician, not an expert in measurement. So this book is not intended to describe subtle aspects of measurement technique. Rather it is intended to explain the various concepts of 'probability' that are applicable in measurement, especially in the production of a justifiable statement of measurement uncertainty. Therefore, where aspects of measurement are described, the emphasis is on ideas that permit the proper use of probability and its ally 'statistics'. There is little more distressing to a scientist than to have the principles of one's discipline misused. So this book is concerned with the correct use of concepts of probability and statistics, and on the legitimate extension of these concepts to address the realities of measurement.

The level at which this book is written is above that of an introductory text. Introductory texts and courses on statistics do not address some of the issues that arise in measurement and often fail to discuss potential areas of confusion for scientists of other disciplines. A simple presentation of basic statistical ideas does not suffice, and for some reason a gulf remains between the understanding of the statistician and that of the typical applied scientist or mathematician. Consequently, this book is written at a higher level, the concepts presented are focused on measurement, and most of the ideas found in introductory statistics books are omitted or assumed. In particular, the emphasis is on reexpressing the relevant principles of data analysis for 'parameter estimation' and on extending these principles to fit the context of measurement.

The physicist, chemist or measurement scientist who reads this book is asked to put aside preconceptions about statistical ideas and terms, especially those reinforced through reading material not written by statisticians. In turn, the statistician who reads this book is asked to engage with the issues that face experimental scientists and to subsequently present the science of statistics in a way that narrows the gulf of understanding (or mis-understanding). Both groups of reader are asked to acknowledge that the language and notation involved must be sufficiently complex to avoid ambiguity. Most importantly, all readers are asked to consider what the actual meaning of an interval of measurement uncertainty should be, especially when this interval is quoted with a figure of 'sureness', say 95%.

That is the central issue raised in this book. It is an issue or question that can be expressed in different forms.

- How can the person for whom a measurement is made properly interpret a 95% interval of measurement uncertainty provided with the result?
- What is the practical implication of the figure 95% in this context?

Readers who are not interested in addressing this question need proceed no further, for the great majority of this book is written to answer it and to develop corresponding means of uncertainty evaluation.

At the forefront of this book is the idea that the meaning of a stated scientific result should be unambiguous. Of course, approximations are unavoidable in modelling and data analysis. But these should be *approximations in number, not approximations in meaning*. It is legitimate to say 'the correct numerical answer is approximately ...' but it does not seem legitimate to say 'the meaning of this answer is approximately ...'. So in this book we require a 95% interval of measurement uncertainty to have an unambiguous and practical meaning. This involves identifying a concept of success in measurement and, by implication, a concept of failure. Without such ideas we have very little to hang our hats upon, very little to claim that we have actually achieved in a measurement, and very little to which we are accountable. Such concepts of achievement and responsibility are not always evident in approaches to the evaluation of uncertainty in measurement.

Structure and style

I trust that this book will be of some benefit to all who engage with it. The book is written in three parts that will be of different value to different readers. Part I describes and explains various principles that provide a basis for the study of measurement uncertainty. It might be lengthier than many readers would expect, but the lack of a set of accepted starting principles is one reason for the confusion and disagreement that we find when approaching this subject. Chapters 1 and 2 consider relevant principles of measurement, in particular the ideas of measurement uncertainty and measurement error. Chapter 3 outlines relevant principles of probability and statistics, explains why the classical approach to statistical analysis is favoured in this book and pays attention to the correct understanding of a *confidence interval*, which is the basic tool of a classical statistician when estimating an unknown parameter using an interval. Chapter 4 considers the idea of treating systematic errors as if they had been drawn from probability distributions, which was an idea recommended by an authoritative group of measurement scientists convened to consider the difficulties of treating these errors. Last, Chapter 5 introduces variant forms of confidence interval that seem particularly applicable in our context of measurement. Many of the ideas developed in Chapters 4 and 5 fall outside the body of classical statistical principles. So the approach taken to the evaluation of measurement uncertainty in this book might be described as one of 'extended classical statistics'.

Part II builds on the principles described in Part I to present methods for the evaluation of uncertainty that are intended to be accurate, meaningful and practical. Chapter 6 completes our preparation for this task by addressing some remaining issues. In particular, it describes a pragmatic step taken to unify the treatment of 'statistical' and 'non-statistical' errors. Chapter 7 discusses the evaluation of measurement uncertainty based on the first-order Taylor's series approximation to the equation defining the 'measurand', the quantity intended to be measured. This chapter describes methodology sufficient for the evaluation of uncertainty in the vast majority of measurement problems. In particular, it puts forward a method that involves the first four moments of the total error distribution, not just the mean and variance. Chapter 8 considers the evaluation of measurement uncertainty when the full non-linear function is retained. In particular, it discusses the use of Monte Carlo simulation of the measurement process. This chapter involves the most complicated mathematical ideas found in this book, and much of it may be omitted. Chapter 9 encourages us to take into account the purpose of making the measurement when considering the appropriate way of representing the measurement uncertainty.

Part III addresses miscellaneous topics. Chapter 10 discusses the measurement of vectors and functions, and Chapter 11 considers the analysis of data from an inter-laboratory comparison conducted to assess, and perhaps refine, a statement of measurement uncertainty. These two chapters largely stand alone, so they may be omitted in a first reading. The later chapters seek to explain and overcome sources of confusion and debate. Readers are asked to study these chapters carefully: the comments contained might be the most helpful found in this book! Chapter 12 considers other approaches to the evaluation of measurement uncertainty, while Chapter 13 focuses on the 'objective Bayesian' approach, which is being advocated in some official documents. Chapter 14 discusses the principles of the *Guide to the Expression of Uncertainty in Measurement* (BIPM et al., 1995; JCGM, 2008a), which is widely seen as an international standard. Last, Chapter 15 discusses the difficult situation where the value of the measurand lies close to a non-physical region into which the interval of measurement uncertainty cannot extend. This situation highlights some of the philosophical issues that motivate the writing of this book.

The text of the book contains sections, examples, definitions and results that are numbered. There are pieces of theory that are numbered and there are also three numbered 'claims'. These items have been written to stand alone. They are numbered within each chapter, and each type of item has its own counter. So Example 3.2 is the second example found in Chapter 3, it is not necessarily found in Section 3.2, and it might precede Definition 3.1.

I have tried to write in a style that is consistent, clear and interesting, and I hope that reading this book will be both informative and enjoyable. I like to think that the book contains new ideas, and that the reader might be surprised by various assertions and so be encouraged to take fresh looks at old problems. Such assertions will be largely associated with two underlying concepts emphasized throughout, these being 'meaning' and 'purpose'.

I have also tried to convey something of my enthusiasm for the classical approach to statistical inference, while not being unduly critical of other points of view. In one or two places – perhaps late in Chapter 8 – this book might be seen as advocating theory that is inapplicable through being impractical. However, some other approaches seem inapplicable through being meaningless or invalid! (It does not seem possible to have an interest in probability and statistics while not having a view of what is meaningful.)

Some simple terms

One theme of this book relates to clarity in notation and terminology. The usage of notation is discussed in some length in Section 3.2, while matters of language and terminology are raised throughout the text. The existence of unambiguous language in science is very important, especially in an interdisciplinary field like this – where ideas of measurement, physics and statistics must meet if progress is to be made. So let me state at the outset the intended meanings of some seemingly simple words. I am convinced that many of the problems we face in this subject arise from different understandings of such words.

- The words *quantity* and *value* will be used in very general senses rather than in any sense that is peculiar to the measurement scientist.
- The word *fixed* will be used to refer to something that cannot change during a process, even though it may be unknown. So a fixed quantity might be regarded as having been set by Providence rather than determined by humankind.
- The verb *to estimate* will mean 'to approximate by some reasonable means'. For something to be approximated, that thing must exist. So for something to be estimated, that thing must exist.[1]
- The verb *to determine* will not be used often. It conveys a much stronger idea than the verb 'to estimate'. To determine the value of something is to find the value exactly. Some scientists speak of 'determining the value of some physical quantity', but I do not understand what they mean. Are they claiming to decree what the value of the quantity is – as in one of the views of measurement discussed in Chapter 1 – or are they using the word 'determine' as others would use the word 'estimate'?
- The word *bias* will be used in a way that conforms to its use in statistics, where the bias is *the error that would be incurred if the estimation procedure were repeated infinitely and the individual results were averaged to form the final result*. This understanding contrasts with the definition of 'measurement bias' given in the *International Vocabulary of Metrology* (VIM, 2012, clause 2.18), which is 'estimate of a **systematic measurement error**', the term in bold being defined in the preceding clause. That definition seems to be an aberration: most scientists would be much more comfortable with the idea that a bias is something to be estimated, not an estimate of something.

This book is written to address the controversies and divisions that exist within this field, some of which are exacerbated by ambiguities about such words. I trust that the book will shed some light on areas in which scientists disagree, especially where there are informal self-taught views of probability and measurement uncertainty. And I would like it to enable the measurement scientist to employ methods acceptable to those concerned about the principles of probability and statistics. In addition, I hope that statisticians reading this book might see the issues that face the measurement scientist and might realize that developing a suitable theory of measurement uncertainty requires thinking outside the statistical square.

---

[1] The claim that we can 'estimate the values of model parameters' when the model and its parameters are known to be abstractions of reality seems to me to be a corruption of English language that has led to much illogical practice.

Part I
Principles

Prudens quaestio dimidium scientiae.
To know what to ask is already to know half.
Roger Bacon (from Aristotle)

To understand God's thoughts we must study statistics, for these are the measure of His purpose.
Florence Nightingale

Truth lies within a little and certain compass, but error is immense.
Henry St John, 1st Viscount Bolingbroke
Reflections upon Exile

... there is nothing so practical as a good theory.
Kurt Lewin
Field Theory in Social Science

1
Foundational ideas in measurement

Our analysis gets under way with the identification of concepts in measurement that involve ideas of chance or probability. We begin this important chapter by observing that there are at least two possible views of measurement but that, for the purposes of accommodating notions of probability, one of these views is to be favoured. There follows a discussion about the existence of a 'true value', which seems to be assumed by those concerned with statistical inference but not by measurement scientists. Then we introduce the ideas of error and uncertainty, and emphasize the importance of being clear about the identity of the quantity that is being measured. Subsequently, we invoke the idea that this quantity may be described by a known function of measurable quantities, which is an idea that is foundational both for our analysis and for analysis in international guidelines. There follows a discussion of the central question: *'What does it mean to associate a specified level of assurance, say 95%, with an interval of measurement uncertainty?'* The chapter finishes with a statement of the basic objective of a measurement procedure.

1.1 What is measurement?

A physical quantity subjected to measurement can be called 'the measured quantity'. Yet the result obtained will often be referred to as 'the measured value'. To 'measure a quantity' seems to imply that we are measuring something that exists before the measurement, but to call the figure obtained 'the measured value' implies that we have measured something that did not previously exist. So is the thing measured a quantity that already exists or is it a value that we bring into being? That is, is measurement the *estimation* of the unknown value of the quantity or is it the *declaration* of the known value of the quantity? These questions might seem silly or unnecessary. If so, the reader might have a fixed idea of 'measurement' and might be assuming that everyone shares that idea. But my experience of interacting with other scientists tells me that such confidence would be misplaced: indeed debates about the nature of measurement have a considerable history (e.g. Stevens (1946)).

The noun 'measurement' is not alone in appearing ambiguous in this way. The term 'judgement' seems to suffer similarly – and it might reasonably be said that measurement involves judging the value of a quantity. When witnesses at a trial are asked to judge the speed at which a car was being driven, they are being asked to estimate the speed of the car. But in pronouncing judgement on a party found guilty, the judge will create or prescribe the punishment. So is judgement – like measurement – the estimation of something or the creation of something?

The *International Vocabulary of Metrology* (VIM, 2012, clause 2.1) defines measurement as a

process of experimentally obtaining one or more **quantity values** that can reasonably be attributed to a **quantity**.

(The words in bold indicate terms defined in that document.) The value attributed to the quantity was not previously associated with that quantity, so this definition is consistent with the idea that measurement is an act of creation or declaration. But it is also consistent with the idea that measurement is an act of estimation. In some situations the value attributed to a quantity can only be understood as an estimate, for example, when measuring a fundamental constant (VIM, 2012, clause 2.11, note 2).

Thus, there seem to be at least two distinct concepts of measurement – one where measurement is the creation or declaration of a known value for the quantity, and the other where measurement is the estimation of an unknown value that is a property of the quantity. Such a fundamental distinction between concepts of measurement will result in differences over the very meaning and role of measurement uncertainty. Therefore this question 'what is measurement?' is addressed at the outset.

Various types of measurement can be defined, but this book is concerned exclusively with the measurement of quantity with a value that exists on a continuous scale, as in the typical measurement problem found in the physical sciences. The view adopted here is that (i) in any well-defined measurement problem of this type we can conceive of a unique unknown value that is the ideal result and that (ii) the act of measurement is the estimation of this unknown value. This view might not be favoured by some readers, but I would ask them to consider the following point, which is argued more fully in the next section: if no such ideal result exists then there is no value to act as a parameter in the statistical analysis. As a result, familiar probabilistic approaches to the analysis of measurement uncertainty become logically invalid.

With this idea of a unique ideal result in mind, I note with gratitude and relief that the *International Vocabulary of Metrology* (VIM, 2012, clause 2.11, note 3) also states that:

When the **definitional uncertainty** associated with the **measurand** is considered to be negligible compared to the other components of the **measurement uncertainty**, the measurand may be considered to have an "essentially unique" true quantity value. This is the approach taken by the ... [*Guide to the Expression of Uncertainty in Measurement*] and associated documents, where the word "true" is considered to be redundant.

The *measurand* is the 'quantity intended to be measured' (VIM, 2012, clause 2.3). Further support is found in the *Guide to the Expression of Uncertainty in Measurement* (the *Guide*) (BIPM et al., 1995; JCGM, 2008a, clause 3.1.2), which states (with cross-references removed):

In general, the **result of a measurement** is only an approximation or **estimate** of the value of the measurand ...

The phrase 'approximation or estimate of *the* value of the measurand' suggests that a unique ideal result exists. As we now argue, this concept seems required if the usual ways of evaluating measurement uncertainty are to be meaningful.

1.2 True values and target values

Typically, a discussion of measurement uncertainty involves a notion of 'probability'. Stating a probability requires the existence of a well-defined event or hypothesis. Consider making a probability statement like '$\Pr(X > \theta + 2\sigma) \approx 0.025$', in which θ represents the value of the quantity being measured and X is the random variable for a measurement result. The event inside the parentheses, '$X > \theta + 2\sigma$', must be well defined if this probability statement is to mean anything: the statement makes no sense if θ does not exist, i.e. if θ does not already have a well-defined, albeit unknown, value. Similarly, consider a statement like '$\Pr(\theta > 2.34) = 0.5$', where θ is treated as a random variable. The event '$\theta > 2.34$' must be well defined if this statement is to mean anything – so, once again, θ must have a well-defined, albeit unknown, value.

A statement like '$\Pr(X > \theta + 2\sigma) \approx 0.025$' features in both the classical and Bayesian approaches to the estimation of θ, while a statement like '$\Pr(\theta > 2.34) = 0.5$' appears in a Bayesian analysis. Therefore, for either of those approaches to parameter estimation to be relevant – which are approaches that result in a *confidence interval* and *credible interval* respectively – there must be some unique unknown value θ that can be seen as the ideal result of measurement (see Rabinovich (2007), p. 605). This can be called the *target value* (Murphy, 1961; Eisenhart, 1963). If the existence of such a value is not accepted then no probability statement involving that value can be made and a different approach to the evaluation of measurement data must be sought. (One possible approach based on using probability as a means of outcome prediction rather than parameter estimation is discussed in Section 12.4. The corresponding probability statements involve potential observations only, like the statement '$\Pr(X > 2.34) = 0.5$'.)

Thus, in the measurement of a physical quantity it is helpful to conceive of a unique target value θ. However, this does not mean that we must accept the existence of a single 'true value' for the quantity. Some scientists seem able to point to situations in which, rather than a measurand having a unique unknown value, there is an unknown *interval* of values consistent with the definition of the measurand. If the unknown limits of this interval are θ' and θ'' then we could set $\theta = (\theta' + \theta'')/2$ and, in so doing, define the target value θ for the measurement, as is illustrated in Figure 1.1(a). And some scientists might even consider there to be an unknown *distribution* of values consistent with the definition of the measurand. In such a case we could consider the target value, θ, to be the median of this distribution, as in Figure 1.1(b). Therefore, the absence of a 'true value' does not preclude us from defining a target value.

Perhaps other scientists encounter situations where the value of the quantity being measured fluctuates with time and is representable as $\theta^*(t)$. As a consequence, the idea of a unique target value or true value might seem inappropriate. But what would be the actual purpose of such a measurement? What aspect of the function $\theta^*(t)$ would be the object of study? If we are interested in estimating the whole function $\theta^*(t)$ then we have a unique *target function* – and we are outside

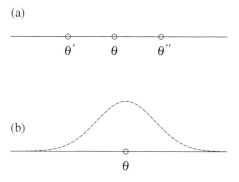

Figure 1.1 Target values θ with two types of 'definitional uncertainty': (a) the target value is the midpoint of an unknown interval; (b) the target value is the median of an unknown distribution.

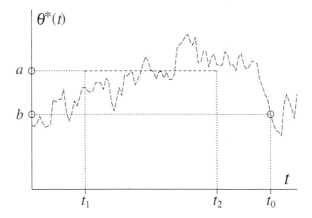

Figure 1.2 Two possible target values when the quantity being measured is fluctuating, $a = (t_2 - t_1)^{-1} \int_{t_1}^{t_2} \theta^*(t)\, dt$ and $b = \theta^*(t_0)$.

the realm of measurement of a scalar, which forms the greater part of this book. However, if we are actually interested in estimating the mean value of $\theta^*(t)$ in some interval $[t_1, t_2]$ then the target value is the scalar $\theta = (t_2 - t_1)^{-1} \int_{t_1}^{t_2} \theta^*(t)\, dt$. And if we are interested in predicting the value of θ^* at some instant t_0 then the target value is the scalar $\theta = \theta^*(t_0)$. These possibilities are illustrated in Figure 1.2.

By such devices, we can define a measurement in a way that the value sought is unique. This enables us to approach the subject of measurement uncertainty using statements of probability that involve this target value: the dispute about the existence or meaning of a 'true value' is put to one side.

1.3 Error and uncertainty

So in any well-defined measurement there is, by some definition, an unknown ideal result of measurement, θ. In that measurement we will obtain a numerical estimate x of θ. If the physical quantity being measured exists on a continuous scale, which is the situation assumed in this book, then θ is unknowable because we cannot measure it with perfect accuracy. Therefore, x cannot be exactly equal to θ, and the quantity $e \equiv x - \theta$ is non-zero, unknown and unknowable. This quantity will feature many times in our analysis, so it requires a name. Although unpopular in some quarters, the usual name for this quantity is the *measurement error*. Thus,

measurement error (e) = measurement estimate (x) − target value (θ).

The quantity x will also be referred to as the *measurement result*; the terms 'measurement result' and 'measurement estimate' will be used interchangeably.

The measurement error is a signed quantity. Its value is unknown, else we would apply a correction and the error would vanish. However, something will be known about its potential magnitude. This leads us to the idea that a statement of *measurement uncertainty* indicates the potential size of the measurement error. It follows that measurement uncertainty is not the same thing as measurement error: uncertainty means doubt, and doubt remains even if the unknown components of error in the measurement happen to cancel.

The idea of a *standard uncertainty of measurement*, or simply a *standard uncertainty*, is one that has been found helpful. That quantity is often denoted u. Suppose that an appropriate figure of standard uncertainty is 0.03 (in some units). The fact that we would write $u = 0.03$, not $u \approx 0.03$, is indicative of the notion that a standard uncertainty is something determined, not approximated. It is a known estimate of something, not something unknown to be estimated. Thus the error is unknowable but the standard uncertainty is known. One definition of standard uncertainty might be

the best estimate of the root-mean-square value of the error in the measurement procedure.

In Chapter 4, we will find that our understanding of the measurement procedure helps us to view the mean error incurred as being zero. The standard uncertainty of measurement then becomes our best estimate of the standard deviation of the underlying distribution of measurement errors. Figure 1.3 illustrates a situation where the standard uncertainty is equal to this standard deviation.

(Figure 1.3 demonstrates a point of presentation in this book. Where appropriate, unknown quantities will be indicated using hollow markers or dashed lines, while known quantities will be indicated using solid markers and solid lines. In this way,

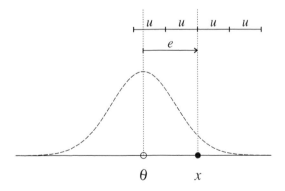

Figure 1.3 Error and standard uncertainty. Dashed line: distribution of potential results when measuring θ; x: known estimate; e: unknown error; u: known standard uncertainty.

1.3 Error and uncertainty 9

we can infer from the body of Figure 1.3 that x is known but that θ and the location of the distribution of potential results are unknown. The same idea is seen at work in Figures 1.1 and 1.2.)

In this book, the concept of a standard uncertainty is not given a great deal of emphasis – and the reader is not obliged to accept the definition of standard uncertainty suggested above. Instead, most of this book is devoted to the task of interpreting and calculating an 'uncertainty interval', which is an output that seems to have greater practical meaning than a standard uncertainty. The concept of an uncertainty interval relates to the idea that, with a specified degree of sureness, we are able to state bounds on the measurement error. Like a standard uncertainty, an uncertainty interval is determined not estimated. This interval depends on the level of sureness we require, and this matter will involve us in a consideration of 'probability'.

The offence of 'error'

For some reason the word 'error' is unpopular in our context. One reason might be distrust of the concept of an ideal result of measurement – without which the terms 'accuracy' and 'error' lack meaning. But, as already indicated, if there is no such thing as an ideal result then the experimentalist has to find an alternative logic when using the concepts of probability.

Another potential reason is that the word 'error' might be thought to imply that a mistake has been made, and the measurement scientist rightly wishes to avoid giving that impression. However, the statistical community finds no offence in the term, and the average citizen does not denigrate market-research companies when they report a 'margin of error' with the result of an opinion poll. Why should it be any different in a measurement problem?

But perhaps the principal cause of the unpopularity of the word 'error' in measurement science is a perception that the techniques of so-called 'error analysis' are inadequate. The techniques of the subject that was known by that name in the middle of the twentieth century do indeed seem limited. The typical text on error analysis would treat all 'random' errors as having arisen from normal distributions, perhaps after invoking the ideas of repeated experimentation and the central limit theorem. More grievously, 'systematic' errors would often be assumed to be either absent or negligible. Thus, the techniques of traditional 'error analysis' are not sufficient for a treatment of the general problem found in measurement science. Yet that is no reason to avoid the concept of error itself. Indeed, the word 'error' seems indispensable: no other word adequately describes the unknown quantity $x - \theta$, and this is a quantity of fundamental importance. Indeed, what is required is a new

Uncertainty about what?

Uncertainty means doubt, but what are we doubtful about? The measurement estimate x is known: there is no doubt about the identity of x. But the target value θ is unknown, so doubt exists about the value (of) θ. Therefore our uncertainty is about θ, not x, and the language used in relation to measurement uncertainty should reflect this. Simple phrases like 'the uncertainty in x' or 'the uncertainty of x' link the concept of uncertainty to x alone, so such phrases do not seem appropriate. The phrase 'the uncertainty that we associate with x as our estimate of θ' seems correct and informative, but is unwieldy. Instead we will write of the *uncertainty of x for θ*. Acknowledging θ as well as x in the statement of uncertainty is not just a matter of principle: we will see in the next section that the same estimate x can have very different amounts of uncertainty associated with it depending on the definition of θ.

This idea of doubt about θ is to be contrasted with the idea of doubt about the next result when measuring θ, which might be the concept of measurement uncertainty assumed by some readers. Such an understanding would seem influenced by a definition like *the standard uncertainty of measurement is the standard deviation of the set of readings that would be obtained in a very large number of observations* (see Dietrich (1973), p. 8). In this definition, there is no mention of any well-defined quantity being measured, no mention of any target value, and no concept of averaging to improve accuracy. Perhaps most tellingly, there is no mention of the possibility of measurement bias. This might be a definition of measurement uncertainty with some appeal to those who do not accept the concept of a 'true value' or 'target value'. As such, it is a definition with potential relevance to material discussed in Section 12.4.

1.4 Identifying the measurand

The quantity being measured must be well defined: the object being estimated must be clear. This is a fundamental idea, and the reason for not addressing it earlier in the chapter is that the concept of measurement uncertainty plays a key role in our discussion.

Imagine a rod that, apart from the existence of some roughness on one end, has the form of a cuboid. This is illustrated in Figure 1.4. The rod stands on the flat end, and 'the length' of the rod is measured by lowering a probe onto the rough face and recording the vertical coordinate of the point of the probe. This is repeated several

1.4 Identifying the measurand

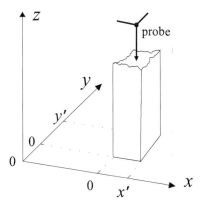

Figure 1.4 Examining lengths of a rod.

times, with the probe being lowered over a different point of the face each time. Let there be n results obtained in total, the jth point on the face being at the coordinates x_j and y_j, and the corresponding vertical distance being $z(x_j, y_j)$. For simplicity, we suppose that this distance is recorded exactly. As might seem reasonable, the arithmetic average of the recorded lengths, $\bar{z} = \sum_{j=1}^{n} z(x_j, y_j)$, is taken as our estimate. But what is the measurement uncertainty to be associated with this figure? The answer depends on the identity of the measurand: what actually is the quantity being estimated?

The figure $\bar{z}$ cannot be called 'our estimate of *the* length of the rod' because the rod has no unique length. It could be said that there are an infinite number of lengths, one for each coordinate (x, y). Yet we have concluded that there is a target value, so something is being estimated and some approximation is being improved by averaging the results of the n measurements. What is being estimated and what approximation is being improved? There are at least two possibilities.

One possibility is that our target is the 'mean length'

$$\mu \equiv \frac{1}{x'y'} \int_0^{y'} \int_0^{x'} z(x, y) \, dx \, dy,$$

which is the volume of the rod divided by the cross-sectional area. This is the mean of the distribution of lengths $z(x, y)$ obtained by associating one length with each infinitesimal element of area $dx \, dy$ on the cross-section. If this distribution of lengths has variance σ^2 and if the coordinates x_j and y_j are chosen randomly then the measurement error $\bar{z} - \mu$ is drawn from a distribution with mean 0 and variance σ^2/n, and the standard uncertainty of measurement will be our best estimate of $\sigma/\sqrt{n}$. Another possibility is that we wish to estimate the length $z_0 \equiv z(x_0, y_0)$ at some unknown point (x_0, y_0) that will subsequently be relevant to our client. Our best estimate of z_0 is again $\bar{z}$, but in this case the error, $\bar{z} - z_0 = (\bar{z} - \mu) + (\mu - z_0)$, is drawn from a distribution with mean 0 and variance $\sigma^2/n + \sigma^2$, so the standard uncertainty of measurement will be our best estimate of $\sigma\sqrt{\{(n+1)/n\}}$.

Therefore – with $\hat{\sigma}$ denoting our estimate of σ – the standard uncertainty of $\bar{z}$ for μ is $\hat{\sigma}/\sqrt{n}$ but the standard uncertainty of $\bar{z}$ for z_0 is $\hat{\sigma}\sqrt{\{(n+1)/n\}}$. The standard uncertainty differs by a factor of $\sqrt{(n+1)}$ depending on whether we wish to estimate the mean length throughout the cross-section or the individual length at an unknown point in the cross-section.

The same idea applies when measuring a fluctuating quantity $\theta^*(t)$, as in Figure 1.2. We might record the value of $\theta^*(t)$ at many instants $t_1, \ldots, t_n$ either to estimate the average level of $\theta^*(t)$ in a well-defined period or to estimate the value of $\theta^*(t)$ at some inaccessible single instant. Suppose the observations are exact, so that the spread in the data reflects only the fluctuation in θ^*. In both cases the measurement estimate will be the sample mean $\sum_{i=1}^{n} \theta^*(t_i)/n$. However, in the first case the standard uncertainty will be $\sqrt{(n+1)}$ times smaller than in the second case.

So the standard uncertainty of measurement is associated not just with the measurement result but also with the identity of the measurand. This is our reason for emphasizing that the target value θ must be acknowledged when quoting a standard uncertainty, as in the phrase 'standard uncertainty of x for θ'.

1.5 The measurand equation

The general measurement problem is more complex than the typical statistical problem of parameter estimation. One reason is that the target value θ is often a known function of many unknown quantities each of which must itself be estimated. This can be expressed as

$$\theta = \mathcal{F}(\theta_1, \ldots, \theta_m), \tag{1.1}$$

where $\mathcal{F}$ is known but each θ_i is unknown. We will call (1.1) the *measurand equation*. Often the quantity physically measured in an experiment is one of the θ_i values, say θ_m. This quantity can be written as

$$\theta_m = \mathcal{G}(\theta, \theta_1, \ldots, \theta_{m-1}), \tag{1.2}$$

and the measurand equation (1.1) is recovered by inversion. We will call (1.2) an *experiment equation*.

Example 1.1 The length of a gauge at 20 °C is measured by relating the length of that gauge to the length of a standard gauge in a comparison conducted at a slightly different temperature. The length of interest l is given by the measurand equation

$$l = \frac{l_s(1 + \alpha_s \lambda) + d}{1 + \alpha \lambda}, \tag{1.3}$$

1.5 The measurand equation

where l_s is the length of the standard gauge at 20 °C, α and α_s are the coefficients of thermal expansion of the gauge and the standard gauge, λ is the difference in the temperature from 20 °C and d is the difference in lengths of the gauges at the experimental temperature. This equation has the form $\theta = \mathcal{F}(\theta_1, \ldots, \theta_5)$ with $\theta_1 = l_s$, $\theta_2 = \alpha_s$, $\theta_3 = \alpha$, $\theta_4 = \lambda$, $\theta_5 = d$ and $\theta = l$. The quantity directly measured in the experiment is d, which is given by the experiment equation

$$d = l(1 + \alpha\lambda) - l_s(1 + \alpha_s\lambda). \tag{1.4}$$

This is an equation of the form $\theta_5 = \mathcal{G}(\theta, \theta_1, \ldots, \theta_4)$, which when re-expressed becomes $\theta = \mathcal{F}(\theta_1, \ldots, \theta_5)$.

This example is based on one found in the *Guide* (BIPM et al., 1995; JCGM, 2008a, H.1). We will revisit this example several times, on each occasion introducing new details in order to illustrate an idea discussed in this book.

In general, some of the θ_i quantities have unknown values that exist before the experiment while others, like those relating to the environment, have unknown values that are created during the experiment. The θ_i quantities with values existing before the experiment can be called *parameters* of the experiment, a parameter of a situation being something that exists outside, and helps define, the situation. Drawing a distinction between the parameters of the experiment and the remaining θ_i values is relevant to the type of 'confidence interval' that we will use as our uncertainty interval for θ. It is also relevant to the evaluation of uncertainty by simulation of the measurement process.

Example 1.1 (continued) The quantities l_s, α_s and α are parameters of the experiment: they existed before the experiment was commenced. In contrast, the unknown values λ and d only exist during the experiment, these values being dependent on the temperature that happened to apply during the experiment.

The measurand equation and experiment equations might be written in a way that involves differences between physical quantities rather than the quantities themselves. This will often facilitate analysis.

Example 1.1 (continued) Henceforth, the two coefficients of thermal expansion will be represented by the coefficient for the standard gauge, α_s, and the difference between the two coefficients, $\delta = \alpha - \alpha_s$. This is because, nominally, the gauges are made of the same material, so the best estimate of δ is zero. Equations (1.3) and (1.4) become

$$l = \frac{l_s(1 + \alpha_s\lambda) + d}{1 + (\alpha_s + \delta)\lambda} \tag{1.5}$$

and
$$d = l\{1 + (\alpha_s + \delta)\lambda\} - l_s(1 + \alpha_s \lambda).$$

So now $\theta_3 = \delta$ and the functions $\mathcal{F}$ and $\mathcal{G}$ are altered slightly.

Each θ_i possesses its own estimate x_i, which will be either generated during the experiment or known before the experiment. The measurement result x and the interval of measurement uncertainty are obtained from the function $\mathcal{F}$, from the estimates $x_1, \ldots, x_m$, and from probabilistic models for the generation of the errors in these estimates. In this book, we will always take the measurement estimate x to be given by

$$x = \mathcal{F}(x_1, \ldots, x_m),$$

and we will call this expression the *estimate equation*. Useful alternative notations for x and x_i are $\hat{\theta}$ and $\hat{\theta}_i$, the caret ˆ being standard statistical notation for 'estimate of'.

Example 1.1 (continued) The estimates of λ and d, which are $\hat{\lambda}$ and $\hat{d}$, are obtained during the experiment. The estimate of δ is known beforehand to be $\hat{\delta} = 0$ and the estimates $\hat{l}_s$ and $\hat{\alpha}_s$ are available from previous work. The estimate of l is

$$\hat{l} = \frac{\hat{l}_s(1 + \hat{\alpha}_s \hat{\lambda}) + \hat{d}}{1 + (\hat{\alpha}_s + \hat{\delta})\hat{\lambda}}.$$

Suppose that $\hat{\lambda} = 0.1\,°C$, $\hat{d} = 254$ nm, $\hat{\delta} = 0$, $\hat{l}_s = 100\,000\,145$ nm and $\hat{\alpha}_s = 11.8 \times 10^{-6}\,°C^{-1}$. Then $\hat{l} = 100\,000\,399$ nm after rounding.

The measurand equation (1.1) will be written as an equality, so that it provides a definition of the target value θ. Some readers might prefer to see it as only an approximation, as merely a model of the real relationship between the quantity of interest θ and the θ_i quantities. If it is an approximation then an analysis of the extent of that approximation should appear in the assessment of measurement uncertainty.

Transferability of uncertainty information

The target value θ in (1.1) might itself become one of the θ_i quantities in a subsequent measurement. So the uncertainty information of x for θ must be expressible in the same form as the uncertainty information of x_i for θ_i. This reproductive property of statements of measurement uncertainty is an important requirement for a general method of uncertainty evaluation.

1.5 The measurand equation

Another desirable property is that of 'associativity'. Ideally, the uncertainty of x for θ will not depend on whether the function $\mathcal{F}$ is applied as in (1.1) or whether it is decomposed into various stages, each of which has its own output. For example, the target value θ in (1.1) might be estimable in two stages according to

$$\theta_0 = \mathcal{F}_1(\theta_1, \ldots, \theta_n),$$
$$\theta = \mathcal{F}_2(\theta_0, \theta_{n+1}, \ldots, \theta_m),$$

so that there is an intermediate quantity θ_0 whose estimate and uncertainty information are inputs to the final analysis for θ. From a practical point of view, we would like the final measurement result and the associated uncertainty information to be independent of the method we choose.

Although desirable, this property of associativity does not seem to be a strict requirement for a scientific system. To mandate that the measurement result and accompanying uncertainty information be the same no matter how $\mathcal{F}$ is decomposed implies that there is a unique 'correct' estimate of θ and a unique 'correct' figure of measurement uncertainty in the problem. As we shall see, the validity of a method of uncertainty evaluation relates to the proportion of measurement problems in which it generates an interval covering the target value. By implication, there is then no unique 'correct' 95% uncertainty interval of x for θ in any *particular* problem. Consequently, there is no scientific imperative for the statement of uncertainty we obtain to be invariant to rearrangement of the measurand equation. Nevertheless, it is worth noting that the analytical methods of uncertainty evaluation proposed in this book do possess the property of associativity.

Looking for the measurand equation

The measurand equation has been introduced before the estimate equation, as if the measurand equation determines the way in which the estimate of θ is obtained. However, there are occasions on which the estimate of θ is determined without a measurand equation being in mind. The measurand equation – which becomes less important – can then be found by working backwards from the estimate equation.

> **Example 1.2** An unknown mass m is to be measured using a device known to have a linear response. The parameters of the response are prone to drift, so measurement of the mass involves recording the outputs of the device when two different standard masses m_1 and m_2 are applied and then recording the output of the device to m. The outputs to m_1 and m_2 establish the estimates of the parameters of the linear response. Subsequently, obtaining an estimate of m is trivial.

The measurand m is not determined by the quantities m_1 and m_2, so the identity of the measurand equation is not obvious. To find it, we consider how the estimate of m will be calculated. Evidently, we will use the equation

$$\hat{m} = \hat{m}_1 + (\hat{m}_2 - \hat{m}_1)\frac{r - r_1}{r_2 - r_1},$$

where $\hat{m}$ is the estimate of m, $\hat{m}_1$ and $\hat{m}_2$ are the stated estimates of the standard masses m_1 and m_2, and r_1, r_2 and r are the readings obtained. An equivalent equation is

$$\hat{m} = \hat{m}_1(1 - f) + \hat{m}_2 f, \qquad f \equiv \frac{r - r_1}{r_2 - r_1}, \qquad (1.6)$$

from which we see that the measurand equation is

$$\theta = \theta_1(1 - \theta_3) + \theta_2 \theta_3,$$

where $\theta = m$, $\theta_1 = m_1$, $\theta_2 = m_2$ and θ_3 is the quantity estimated by the ratio f, i.e. $\theta_3 = (m - m_1)/(m_2 - m_1)$. The corresponding estimate equation is (1.6).

The model of the measurement

The measurand equation has elsewhere been called the 'measurement model' (JCGM, 2008b). This is regrettable, because measurement is a process that incurs error, and any model of a measurement must contain a description of how this error arises.

As a noun, the word 'model' means different things to different people. It is used in this book to mean *an approximate description of an aspect of reality, with this description being developed for a specific purpose*. Any reference to a (mathematical and statistical) model of the measurement as a whole should be to a body of equations comprising the measurand equation (1.1), underlying experiment equations like (1.2), and equations for the generation of appropriate θ_i values and the generation of the errors $e_1, \ldots, e_m$, where $e_i = x_i - \theta_i$. Such a model will be fit for our purpose of obtaining a measurement result and a statement of measurement uncertainty. We will consider the explicit statement of such a model in Chapter 6.

1.6 Success – and the meaning of a '95% uncertainty interval'

It is now appropriate to ask the basic question underlying the controversies about error analysis and uncertainty analysis. *What is to be understood when a specified level of reliability, e.g. 95%, is attached to an interval of measurement uncertainty?* Once the answer to this question is known, we can look for an interval that obeys that understanding. (And if no practical answer is available then perhaps we should

1.6 Success – and the meaning of a '95% uncertainty interval'

not report any interval at all!) For simplicity, we will work throughout with the familiar level of 95%. So the question is *'When a "95% interval of measurement uncertainty" or "95% uncertainty interval" is quoted at the end of a measurement, how can the recipient of the interval legitimately interpret it?'* This is the most important question to be addressed by readers of this book. I hope that the discussion now given and the material presented in the following chapters are sufficient to provide a clear and persuasive answer.

Despite the fact that a practical interpretation is required, we put forward a definition of a 95% uncertainty interval that, at this stage in the book, seems rather vague.

> **Definition 1.1** A *95% interval of uncertainty* for the target value θ is an interval that we can be 95% sure contains θ.

This definition alludes to the fact that the role of an interval of measurement uncertainty is to contain the target value. An interval that does so, and the measurement process that produced this interval, will be called *successful*. And the event of obtaining an interval containing the target value will be called *success*. So the definition speaks of an interval in which 'we can be 95% sure' of having been successful.

Definition 1.1 may be compared with the definition of a 95% coverage interval implied in the *International Vocabulary of Metrology* (VIM, 2012, clause 2.36), which is

an interval containing the set of true quantity values of a measurand with probability 95%, based on the information available.

This definition uses the word 'probability', which seems to be interpreted differently by different people (as we shall see in the next chapter). Definition 1.1 deliberately avoids any reference to probability. Despite that, the calculus of probability is certainly relevant to the task of calculating a 95% uncertainty interval – and much of this book is devoted to describing relevant concepts of probability and statistics. The foundations of this analysis are described in Chapter 3, where we examine the idea of probability and also examine the idea of sureness – thus giving clarity to Definition 1.1.

Let us now be more specific. The figure of 95% in Definition 1.1 must relate to some level of performance, else this figure would seem to have no practical meaning. The interval itself is either successful or unsuccessful, so its level of performance is either '1' or '0'. Therefore, the figure 95% does not relate to the interval itself but instead relates to the procedure that was used to calculate the interval. It can only be some kind of success rate: there seems to be no other practical interpretation. The figure of 95% must be equal to, or be a lower bound on, the proportion of intervals obtained that are successful. We therefore state one basic idea on which the majority of this book rests:

> **Claim 1.1** Our methods of analysis should be designed so that at least 95% of all '95% uncertainty intervals' will contain the corresponding target values.

Notwithstanding other concerns like interval-length, it is acceptable to exceed the advertised success rate. Hence the figure of 95% is to be seen as a minimum success rate, not an exact success rate. This idea is consistent with common uses of high figures of sureness. For example, when I say 'I will take a course of action if I am 95% sure of success' I actually mean 95% or more.

Furthermore, the following claim seems reasonable.

> **Claim 1.2** We can be at least 95% sure that a particular interval contains θ if it was created by a procedure known to be successful on 95% of occasions and if there is no reason to regard the interval as being less reliable than any other that might have been obtained.

Therefore, with the help of Definition 1.1, we can also make a third claim.

> **Claim 1.3** Any interval obtained in a measurement with target value θ can be called a '95% uncertainty interval for θ' if it is no less reliable than the typical member of a hypothetical set of intervals for many different measurements of the same type, at least 95% of which would be successful.

Claim 1.3 forms the basis of the means of evaluating intervals of measurement uncertainty described in Part II. It involves the ideas of:

1. success in measurement, success being the generation of an interval containing the target value,
2. a hypothetical large set of measurement problems, each of which admits the potential calculation of an interval for the corresponding target value,
3. an interval being typical, by which we mean that there is no external knowledge as to why this interval should be seen as less reliable than others in the set.

Definition 1.1 and Claim 1.3 may be contrasted with the definition of a 95% interval of measurement uncertainty implied in the *Guide* (BIPM et al., 1995; JCGM, 2008a, e.g. clause 2.3.5), which is

an interval about the result of a measurement that may be expected to encompass 95% of the distribution of values that could reasonably be attributed to the measurand.

This definition makes no reference to any idea of success in measurement and does not hold the method of analysis accountable to any performance criterion. It includes the rather vague phrase 'reasonably be attributed', which we consider in Section 12.4, and it leaves the reader wondering whether it is a *value* or a *distribution* that is being attributed to the measurand. It also alludes to the questionable idea that values attributable to a measurand fall in a distribution rather than a range.

Claim 1.3 involves the concept of a 'typical' member of a set of intervals. However, there are some measurement situations where the idea that an interval is typical is problematic, the most obvious of these situations being where the analysis procedure might generate an interval that extends into a non-physical region. So this claim does not act as a general definition of a 95% uncertainty interval. Rather, it tells us that an interval obtained using a procedure with certain properties qualifies to be called a 95% interval of measurement uncertainty. Thus, we have identified an important kind of interval that can be described as a '95% uncertainty interval' but have not fully defined that term. Definition 1.1 will only gain greater clarity when we consider the ideas of sureness and probability in Chapter 3.

1.7 The goal of a measurement procedure

This chapter has outlined ideas of measurement that involve the concept of chance or probability. Its foundation is the idea that in every measurement there is a 'target value', which can be seen as the result that would be obtained under ideal circumstances. A consequential idea is that of 'success' in measurement, by which we mean the generation of a known interval containing the target value. A third idea, which we have not described until now, is the idea that a short successful interval is of greater benefit than a long successful interval because it corresponds to a more precise statement about the target value. These ideas will now be combined to give a succinct definition of the goal or objective of a measurement.

Suppose we measure the mass m of an object and obtain the estimate 10.030 kg. (So $\theta = m$ and $x = 10.030$ kg.) The output of the procedure must be some kind of assertion about the target value. One possible assertion would be '$m \approx 10.03$ kg', but this statement would be of dubious value without the degree of approximation also being communicated. The degree of approximation might be expressed as a standard uncertainty, but in this book we place more emphasis on the expression of an interval of uncertainty. If the measurement procedure incurs error with mean zero and standard deviation 0.009 kg and if an interval of 95% measurement uncertainty is required then the interval stated for m would be [10.012, 10.048]. In effect, to report this interval is to make the assertion

'the mass of the object is between 10.012 kg and 10.048 kg' (1.7) and to indicate that there is a 95% assurance or trustworthiness associated with it. Such an assertion is either correct or incorrect, and we cannot know which without further examination: that is a fact of life. But (usually) we can put in place a method of analysis that leads us to have 95% assurance in its correctness. In doing so, we acknowledge that 1 in 20 of the assertions that we make will be false. It is clear that, if correct, statement (1.7) is more informative and useful than a statement involving limits separated more widely.

With these things in mind, we address a question that seems to be ignored all too often: what is the goal or objective of a measurement procedure? The answer suggested is as follows.

The objective of a measurement is to make a correct and useful assertion about the value of the measurand, more specifically, about the target value.

The rest of the book is concerned with describing methods of reasoning, methods of modelling and methods of data analysis by which we can achieve this objective – but only 19 times out of 20!

2
Components of error or uncertainty

As described in Claim 1.1, our methods of analysis are to be designed so that at least 95% of all '95% uncertainty intervals' will contain the corresponding target values. Consequently, it is important to understand the natures of the errors in our measurement processes. In this section, we outline several sources of error and pay particular attention to the statistician's concept of sampling error. On its own, this concept is inadequate for the proper treatment of a measurement problem: a logical treatment of measurement uncertainty must involve other kinds of error.

Before proceeding, let us briefly examine the idea of 'measurement error' in a statistical context, so that we will not confuse it with statistical error in a measurement context. In a statistical context, the basic task is to infer something about a whole population from examination of a finite sample. Typically, the examination of each element in the sample is seen as exact, so the resulting error is taken to arise entirely from the finiteness of the sample. This is called *sampling error*. For the statistician, an idea of 'measurement error' is secondary and might be invoked when the examination of an element in the sample cannot be regarded as exact. Perhaps the idea of 'measurement error' most often encountered in statistical literature will be that of studying the population of responses at an erroneous setting of a factor defining the population, e.g. an erroneous x-value in a 'y on x' regression model (e.g. Fuller (1987), Eq. 1.1.2). So there is an issue of terminology here. To a statistician the term 'measurement error' refers to a secondary source of error. But that will not be so for readers of this book, in whose context measurement error contains statistical error, and not vice versa.

2.1 A limitation of classical statistics

The phrase 'classical statistics' was used in Chapter 1 without explanation. It refers to the paradigm of frequentist statistical inference, where a probability is interpreted as a relative frequency and where concepts such as bias, mean-square error, minimum-variance estimation and confidence level are relevant. The word 'classical' correctly suggests that this is the paradigm of statistics likely to have featured most strongly in the reader's education – even though it might be argued that the Bayesian paradigm arose earlier in the history of statistics.

The fundamental concern of classical statistics is the quantification of the potential error when a finite sample is used to represent a population. At first glance, this concept has nothing to do with issues of measurement. However, inspection shows it to be related to two basic problems: the measurement of a fixed quantity by repeated observation and the calculation of a calibration curve by regression. In the problem of repeated observation, the observations, or indications, are regarded as a sample from the population that would exist if the repetition could be continued indefinitely: for example, Figure 2.1 depicts a sample of seven indications from a potential population. In the calibration problem, the responses of a system are observed at several predetermined input values. The response y_i at any input value x_i is regarded as a sample of size 1 from the hypothetical population of responses that would be obtained if the system were repeatedly studied at x_i. This is depicted in Figure 2.2.

So the essential principle of classical statistics is applicable in some important elementary problems in measurement, and the associated statistical theory furnishes solutions. However, the basic principles of classical statistics seem insufficient to accommodate the subtleties of general measurement problems. For example, in statistics the quantity studied is fixed and repetition leads to different estimates of it, but sometimes in measurement the estimate is fixed and the quantity measured takes different values instead (as we shall see in the next section). More importantly, the idea of a 'systematic' error, which is fundamental in measurement, features little in basic classical statistics. So there are a number of ideas that must somehow be taken into account if classical statistics is to be applied in complex measurement problems. In particular, we suggest that the concept of 'repetition of the procedure', which gives meaning to the idea of probability, must be broadened. Without our concepts of 'repetition' and 'procedure' being able to accommodate

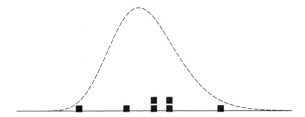

Figure 2.1 One idea of sampling in measurement: a sample of indications from an unknown or theorized population.

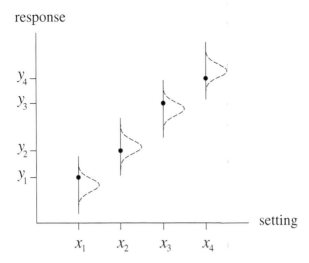

Figure 2.2 A second idea of sampling in measurement: raw data $\{(x_i, y_i)\}$ in a basic calibration procedure. Each x_i value is a fixed setting and each y_i value is regarded as a single element drawn from a distribution that would be revealed if the calibration could be repeated indefinitely.

systematic errors, the principles of classical statistics seem inadequate. This idea is addressed in Chapter 4.

Therefore, classical statisticians must think outside their usual context if their techniques are to have great relevance to the typical problem in measurement science. Despite this, the statistical concept of error remains important in measurement, and it is usually discussed before other sources of error.

2.2 Sources of error in measurement

We now consider various means by which an estimate, x_i, differs from the quantity that it estimates, θ_i. That is, we consider various possible sources for an error $e_i = x_i - \theta_i$. Through the measurand equation (1.1), the e_i errors combine to form the overall measurement error $e = x - \theta$.

Statistical error

Statistical error is error that exists because we have a finite sample of experimental data from a potentially infinite population. The most common type of statistical error in measurement is associated with the idea that an observation or indication is drawn from some distribution located around the relevant target value. The difference between the observation and the target value can be understood to be the sum of errors from influences that are too small and too numerous to be of individual concern. By the marvellous phenomenon described in the central limit theorem for non-identical variables (see Section 3.5) the error becomes well modelled as following a Gaussian (normal) distribution when the experiment is repeated. The idea of not knowing or not caring where this error comes from is associated with the term 'pure error' sometimes used in statistics.[1]

A second type of statistical error arises when estimating a rate in a stochastic process, such as a background rate of radioactive decay. The statistical model in this case usually involves the Poisson distribution. Again this error is caused by the finiteness of our data. The exact result would be found if we were able to measure over an infinite amount of time; the error only arises because this is not possible.

A third type of statistical error is associated with the idea that a specimen examined is not representative of the whole. This 'inhomogeneity error' will be particularly relevant in chemistry and biology. (In those fields, a specimen seems to be called a 'sample'. There is potential for confusion between *this* idea of a sample, which relates to a single specimen from some continuous bulk, and the statistical idea of a sample comprising several discrete elements from a population. This is a reason for preferring the term 'inhomogeneity error' to 'sampling error' here. Life would be so much simpler for statisticians who are seeking to engage with applied scientists – and consequently these statisticians could be much more help – if chemists and biologists used the word 'specimen' instead of 'sample'.)

Calibration error

Most measurement situations involve the use of equipment that has been calibrated in some way, either by the experimenter or by the equipment manufacturer. Random influences during that process lead to the calibration curve obtained being inexact. The calibration might have involved a finite number of repeated observations, so the error $x_i - \theta_i$ might initially have been seen as a statistical error. However, when the calibration is completed, the error in the result at a certain setting becomes an unknown error that remains constant through all subsequent uses of the apparatus at that setting. The error ceases to be statistical in nature (Kirkup and Frenkel, 2006, pp. 44,45).

The simplest example of a calibration error is the error incurred when accepting the reading on a device even though the manufacturer states that the reading is only correct to within a specified amount. If the manufacturer specifies that the reading on the device is accurate to within 1 unit then (i) it will be assumed that the error does not exceed 1 in magnitude and (ii) it will usually be assumed that the error remains constant when the device is used at any fixed setting.

---

[1] See Alfassi *et al.* (2005, pp. 48,49) for a thoughtful decomposition of the set of components of influence in a measurement.

Errors in reference values

The adoption of reference values obtained in previous experiments is analogous to the use of calibrated equipment. The error incurred in forming a reference value x_i might have been statistical during the previous experiment, but the error incurred when subsequently using the reference value is a constant. The same principle applies when x_i is an appropriate estimate of a fundamental constant.

Matching error

In the types of error described so far, the estimate x_i is created in some process that had a random component. Now we consider a situation where the estimate is fixed and cannot be seen as the outcome of a random process.

Often the relationship between physical quantities is controlled imperfectly, usually when quantities are not perfectly matched. The corresponding error may be called *matching error*. An example of this is found in Example 1.1.

Example 1.1 (continued) Recall that the length of a gauge at 20 °C is given by

$$l = \frac{l_s(1 + \alpha_s\lambda) + d}{1 + (\alpha_s + \delta)\lambda},$$

where δ is the difference in the coefficients of thermal expansion of the gauge and the standard gauge. Because the gauges are nominally made of the same material, the estimate of δ is the fixed value 0, i.e. $\hat{\delta} = 0$. This estimate would not change when repeating the experiment, so the error in this estimate, which is a matching error, cannot traditionally be thought of as the outcome of a random process.

Discretization error

Discretization error (or digitization error) arises when there is a finite unit of resolution in the recording device or display. The size of the error depends on whether the device rounds to the nearest unit or truncates. If x'_i is the undiscretized input to the device and if the device rounds to the nearest integer then the recorded figure is

$$x_i = \lfloor x'_i + 0.5 \rfloor,$$

where $\lfloor \cdot \rfloor$ means rounding *down* to the next integer. The quantity $x_i - x'_i$, which is the resulting component of error, depends on how much the input x'_i differs from the nearest unit. Consequently, this depends on the input in a way that is not monotonic, and this means that the uncertainty to be associated with discretization error becomes difficult to treat in a way that is theoretically satisfactory (Willink, 2007b). We will find in Section 5.1 that an idea of 'average confidence' is helpful to overcome this problem. It is argued there that the value of the input can be treated as if it were randomly placed between neighbouring discrete points on the scale. This simplifies the task of including a component of uncertainty associated with discretization.

Fluctuation in environmental factors

Suppose that a measurement involving the collection of a set of data is conducted in the presence of a fluctuating environmental variable, for example temperature. A component of error will arise when this fluctuating variable is represented in the analysis by a single figure. The following analysis suggests that this component is, however, largely contained within the total statistical error.

If one of the θ_i quantities in the measurand equation $\theta = \mathcal{F}(\theta_1, \ldots, \theta_m)$ is a fluctuating environmental variable then, for the unique target value θ to exist, at least one other θ_i value must be fluctuating. This will usually be the quantity physically measured, say θ_m, which will be given by the experiment equation $\theta_m = \mathcal{G}(\theta, \theta_1, \ldots, \theta_{m-1})$. Let the fluctuating environmental quantity be θ_1. The experiment equation becomes

$$\theta_m(t) = \mathcal{G}\left(\theta, \theta_1(t), \theta_2, \ldots, \theta_{m-1}\right),$$

which shows the dependence of θ_1 and θ_m on time. If the data arising in the measurement of θ_m are n observations obtained at different times $t_1, \ldots, t_n$ then these observations are, in fact, individual estimates of n different quantities, $\theta_m(t_1), \ldots, \theta_m(t_n)$. So the measurand θ satisfies the n simultaneous measurand equations

$$\theta = \mathcal{F}\left(\theta_1(t_i), \theta_2, \ldots, \theta_{m-1}, \theta_m(t_i)\right), \qquad i = 1, \ldots, n.$$

Let x_m be the estimate of θ_m formed from the n observations. It would be common to take the quantity $x = \mathcal{F}(x_1, \ldots, x_m)$ as the measurement estimate but to evaluate the uncertainty of x for θ using the basic equation $\theta = \mathcal{F}(\theta_1, \ldots, \theta_m)$ without taking into account the time dependence. In that case the spread in the observations, some of which will be due to fluctuations in θ_m, will be treated as an entirely statistical phenomenon, and the variability due to the environmental fluctuation will be seen as pure error.

Example 1.1 (a variation) The quantity physically measured in Example 1.1 is the length difference

$$d = l\{1 + (\alpha_s + \delta)\lambda\} - l_s(1 + \alpha_s \lambda),$$

2.2 Sources of error in measurement

where λ is the difference in the temperature from 20 °C. Let $t_1, \ldots, t_n$ be times during a period in which the temperature was fluctuating. So

$$d(t_i) = l\{1 + (\alpha_s + \delta)\lambda(t_i)\} - l_s\{1 + \alpha_s\lambda(t_i)\}, \qquad i = 1, \ldots, n.$$

Imagine that the lengths of the gauges are compared at each of these times and that $\hat{d}(t_i)$ is the resulting estimate of $d(t_i)$. If we invoke the measurand equation

$$l = \frac{l_s(1 + \alpha_s\lambda) + d}{1 + (\alpha_s + \delta)\lambda},$$

and see each unknown in this equation as having a unique value then, presumably, d and λ in this equation will be defined to be the average values of $d(t)$ and $\lambda(t)$ during the period of measurement. The effect of the fluctuation of temperature will appear as spread in the $\hat{d}(t_i)$ data and the estimate of d will be $\hat{d} = \sum_{i=1}^{n} \hat{d}(t_i)/n$. The resulting component of error in this estimate of d will be incorporated into the statistical error.

So (if averages and means are used) there seems to be no need to include a component of uncertainty related to the unmeasured fluctuation in an environmental factor during a period of data acquisition. This fluctuation will be reflected in the spread of results, where it will simply be one of the sources of the pure error.

Errors linked to the measurand equation

Another cause of error is the local approximation of the measurand equation by a linear equation. The function $\mathcal{F}(\theta_1, \ldots, \theta_m)$ near the known point $(x_1, \ldots, x_m)$ is usually representable by a truncated Taylor's series. The corresponding linear approximation is

$$\mathcal{F}(\theta_1, \ldots, \theta_m) \approx \mathcal{F}(x_1, \ldots, x_m) - \sum_{i=1}^{m}(x_i - \theta_i)\frac{\partial \mathcal{F}}{\partial x_i}.$$

(See Chapter 7.) The difference between the quantities on each side of this equation might be called *linearization error*. This is illustrated in Figure 2.3 for a situation with $m = 1$.

As implied in Section 1.5, the measurand equation (1.1) might sometimes be regarded as an approximation better written as $\theta \approx \mathcal{F}(\theta_1, \ldots, \theta_m)$. If this is the case then the small quantity $\mathcal{F}(\theta_1, \ldots, \theta_m) - \theta$ can be called a *specification error*.[2]

---

[2] The common term 'misspecification error' seems to be a misnomer. All models are inexact, so the error is incurred simply by specifying the model!

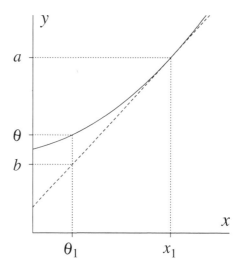

Figure 2.3 An example of linearization error with the measurand equation $\theta = \mathcal{F}(\theta_1)$. Solid line: the function $y = \mathcal{F}(x)$. Dashed line: the approximating function $y = \mathcal{F}(x_1) - (x_1 - x)\partial \mathcal{F}/\partial x_1$. Also $a = \mathcal{F}(x_1)$ and $b = \mathcal{F}(x_1) - (x_1 - \theta_1)\partial \mathcal{F}/\partial x_1$. The linearization error is $b - \theta$.

2.3 Categorization by time-scale and by information

The previous section has described different sources of error in measurement. Now we discuss the idea that the corresponding components of error can be categorized according to the way that measurements made at different times are affected by them or according to the information that we have about them.

Traditionally, errors are understood to be of two sorts – random and systematic. A *random error* is an error that would take an unrelated value if the experimental part of the measurement were repeated. Thus, statistical errors are random errors. Some matching errors will also be random errors, because the matching of quantities might sometimes be seen as part of the experiment. We can stipulate that the average value of a random error is zero because any non-zero mean can be interpreted as a 'fixed error', as shall be defined below. In contrast, a *systematic error* is a component of error that in replicate measurements stays constant or changes in a predictable way (VIM, 2012, clause 2.17).[3]

A useful equation describing the relationship between an estimate x and its target value θ is

$$x = \theta + e_{\text{sys}} + e_{\text{ran}}, \qquad (2.1)$$

where e_{sys} and e_{ran} are the total errors from systematic and random components respectively. Often this equation is used when all systematic components are

---

[3] Some authors have used the term 'systematic error' to refer only to a component of error that remains constant (e.g. Beers (1957), Mandel (1964)).

2.3 Categorization by time-scale and by information

constant – so it is helpful to draw a distinction between constant components and those that vary predictably. Therefore, an error that takes the same value whenever we carry out the experiment (during some relevant period of time) will be called *fixed* and an error that varies predictably will be called *moving*. Some examples of fixed errors are errors of calibration, errors in reference values and errors in estimates of fundamental constants. The best example of a moving error is perhaps an error that is dependent on temperature when the temperature is uncontrolled and unmeasured. Thus, a systematic error is either a fixed error or a moving error. So, in the commonly assumed situation where all systematic errors are unchanging, (2.1) seems better written as

$$x = \theta + e_{\text{fix}} + e_{\text{ran}}, \quad (2.2)$$

with the implication being that there are no moving errors.

Example 2.1 A device is calibrated and then repeatedly used to measure a mass m. The calibration is imperfect and an error e_{cal} remains. This is a fixed error because it applies in every measurement of m. The estimate of m obtained in the ith measurement is modelled as

$$\hat{m}_i = m + e_{\text{cal}} + e_{\text{ran},i} + \delta,$$

where $e_{\text{ran},i}$ is a random error and δ is a fixed error. There is no relationship between $e_{\text{ran},i}$ and $e_{\text{ran},j}$ for $i \neq j$ except for the fact that they are drawn from the same distribution, which is a distribution taken to have mean zero because any non-zero mean makes up δ.

The traditional classification of errors as random or systematic is useful for many purposes, but it seems incomplete. For example, a discretization error is neither random nor fixed, nor is it the sum of a random error and a fixed error. This error is fixed if every other component of error is fixed, but it will contain a random part otherwise. So it does not fit our description of a moving error either. Therefore, discretization error is difficult to deal with logically. Thankfully, the effect of discretization error is often negligible (Willink, 2007b). Furthermore, we can treat such an error by appealing to the idea of 'average confidence' discussed in Section 5.1.

The classification of errors as random, fixed or moving is a classification according to the length of time for which the error persists. Random errors persist on an infinitesimal time-scale, fixed errors persist on an infinite time-scale and moving errors can be understood as persisting on intermediate time-scales. In contrast, the approach to the evaluation of uncertainty described in the *Guide* (BIPM et al., 1995; JCGM, 2008a) is based on a different classification – with evaluation of a component of uncertainty being described as 'Type A' if it is carried out by statistical means and 'Type B' otherwise. Likewise, we shall define an error to be a *Type A* error if it is statistical and to be a *Type B* error otherwise. It can be seen that this classification relates to the way that the scale of the error is assessed. The scale of a Type A error is estimated from experimental data, for example a sample standard deviation, but the scale of a Type B error is decided from other considerations, such as an equipment manufacturer's statement of accuracy.

> **Example 2.1 (continued)** Many measurements are again made of the mass m, but now the device is recalibrated at various times during these measurements. The estimate of m obtained in the ith measurement is now modelled as
>
> $$\hat{m}_i = m + e_{\text{cal},i} + e_{\text{ran},i} + \delta,$$
>
> where $e_{\text{cal},i}$ is the calibration error applicable for the ith measurement. Each $e_{\text{ran},i}$ is independently drawn from the same distribution, and the standard deviation of this distribution will be estimated from the measurement results. So $e_{\text{ran},i}$ is a Type A error. In contrast, $e_{\text{cal},i}$ is a Type B error because its potential size is assessed from knowledge of the calibration process. The error δ might be of Type A or Type B.
>
> (Here $e_{\text{ran},i}$ is still a random error and δ is still a fixed error. However, $e_{\text{cal},i}$ is not a fixed error because it does not take a single value throughout the whole process. Neither is it a random error because the same calibration error endures for several measurements.)

In many situations, there will be a one-to-one correspondence between random errors and Type A errors and between systematic errors and Type B errors. However, this will not always be the case. In particular, it is possible to conceive of a random error that is drawn from a distribution whose standard deviation is assessed by expert opinion. That is, we can envisage a random error of Type B.

The classification of errors as random, moving or fixed is based on the longevity of the value of the error, while the classification of errors as Type A or Type B is based on the type of information available about the error. Distinguishing between random and fixed errors is important in many practical situations, some of which are described in Chapter 9, while distinguishing between Type A and Type B errors is relevant when justifying a method of evaluating measurement uncertainty.

2.4 Remarks

In summary, we can say that many measurement problems possess features that are not properly treated using standard methods of classical statistics. In particular, the existence of errors that are not sampling errors complicates these problems. We require a methodology that enables all errors to be combined in a logical manner. The following chapters take various steps towards the development of such a methodology. Chapter 3 discusses ideas of probability and statistics in some depth. In effect, it addresses the question, 'what are methods for the evaluation of uncertainty to achieve in real terms and what concept of "probability" should be adopted to facilitate this?' Chapter 4 describes how the answer to that question permits redefinition and reinterpretation of the idea of repeating a measurement, which is an idea that is at the heart of the difference between random and systematic errors. Subsequently, Part II describes methodology developed with these principles in mind.

The meanings of 'random'

Before proceeding, it is appropriate to make a point about language. The word 'random' seems to be used in different senses by non-statisticians and statisticians. In accordance with our description above, many non-statisticians would call the quantity $e_{\text{ran},i}$ in Example 2.3 a random error. However, the strict statistician or probabilist would see $e_{\text{ran},i}$ as being the *realization* or *outcome* of a random variable, and therefore as being random no more! This difficulty – which is also related to a common usage of the term 'confidence interval' – is commented on further in Section 3.2. The significance of this issue should not be underestimated, for the main division between statistical camps relates to what can be regarded as random in the statistical sense. So any misunderstanding of this idea can only hinder appreciation of the debate.

In keeping with the statistical understanding of the word 'random', which will feature significantly in the rest of this book, we now tend to refer to an unknown numerical error like $e_{\text{ran},i}$ not as a random error but as the realization or outcome of a random error.

3
Foundational ideas in probability and statistics

I used to think that statistics was a branch of mathematics, but I have come to believe that idea to be incorrect. There are ideas of statistical inference that fall outside the realm of mathematics and which require additional understanding. The basic difficulty seems to be a lack of clarity about the meaning of the word 'probability' and a lack of awareness of this problem. Many people do not realize that there is controversy about probability or that this issue is more philosophical than mathematical. Is probability only the long-run relative frequency of an outcome in a repeatable experiment, as a classical, frequentist, statistician might hold, or is probability really just a strength of belief, as a Bayesian statistician might argue? Those unaware of the division between these views can easily mix the associated ideas in ways that do not stand up to logical examination, while others might reasonably conclude that two factions are arguing over possession of one word.

One starting point for the analysis in this book is the claim that probability must be a practical thing. The assignment of a certain numerical probability to an event or hypothesis must have a practical meaning if the calculus of probability is to be the basis of a system said to have practical value. So we look for an outworking of the idea of probability in real and practical terms.

In the next section, we address this subject by first agreeing with the Bayesian statistician's claim that probability is a 'degree of belief' and then arguing that this degree of belief must rely on a concept of long-run behaviour, as in frequentist statistics. We show that the actual difference between the two main points of view is not in the meaning of probability but in the range of events seen as legitimate subjects of statements of probability. This goes a small way to bridging the division between the two camps. The understanding of probability that is put forward then enables us to advocate the frequentist ideas of confidence and confidence intervals.

3.1 Probability and sureness

Let H denote the event or hypothesis that the word 'supersede' is spelled correctly here. How do you express any level of doubt that you might have about the truth of H? You might say something like 'I am 95% sure that H is true' or 'I am 95% sure of H', but what would be your meaning in practical terms? What does the statement 'I am 95% sure' actually mean?

One point to be made is that the figure of 95% must mean something, else the statement can be replaced by the statement 'I am 75% sure' – and this is not the case. Another important idea is that this statement is personal and subjective. The idea that you are 95% sure of H does not involve anyone else's view of H. The statement simply refers to *your* state of mind. So we require a definition of a specific level of *your* sureness if the statement 'I am 95% sure that H is true' is going to convey what you mean to your listener.

A suitable definition of your level of sureness might be as follows.

Imagine that the truth or falsehood of a hypothesis H will be determined, that p_0 is a specified real number in the interval [0, 1) and that you are invited to accept one of the following two contracts: (a) you will receive a prize if a decimal fraction to be randomly chosen turns out to be less than p_0, and (b) you will receive the same prize if it transpires that H is true. The level of sureness that you attribute to H is less than p_0 if you favour contract (a), greater than p_0 if you favour contract (b), and equal to p_0 if you favour neither.

And a similar definition is:

Imagine that the truth or falsehood of H will be determined. The level of sureness p that you attribute to H is the maximum number of dollars you would pay to receive 1 dollar if it transpires that H is true.

Both of these definitions implicitly require you to be unconcerned about any cost to the other party. The second definition also involves the idea that you might incur a net loss of p dollars, which is money that could be sufficient to cover a small but urgent bill. So the second definition requires you to ignore the concept of the 'utility' of money. The first definition does not require you to ignore this utility, so it seems to be superior. Notwithstanding this, the two definitions are functionally equivalent.

The emphasis on gambling for your personal gain in these definitions seems regrettable. So, to avoid further insult, we will replace you by a hypothetical person acting in a consistent fashion and out of unashamed self-interest – the adjective 'consistent' indicating that the person always bets according to his or her belief and never according to his or her emotions. Such a person is sometimes called a 'rational agent', but we will use the term 'consistent gambler', the rationality of always acting out of self-interest being debatable. Moreover, we will then replace the idea of this person's level of sureness in H by the idea of the 'probability' that he or she attributes to H. So our definition of personal (or subjective) probability is:

> **Definition 3.1** The truth or falsehood of a hypothesis H will be determined. The *probability* that a consistent gambler attributes to H is the maximum amount he or she would pay in order to receive 1 unit if H is found to be true. This probability is the *degree of belief* of the gambler in the truth of H.

The unit is understood to be small enough for the concept of utility to be irrelevant.

Long-run behaviour

So personal probability can be defined in terms of the theoretical behaviour of a person entering into a contract or bet. But what would be driving that behaviour? In effect, the definition is incomplete without this being known. Definition 3.1 implies that if the price is right then a person will enter a bet even if his or her personal probability is less than 0.5, in which case the person thinks it is more likely he or she will lose than win! Why would the person do that? The answer is that he or she regards the *mathematical expectation* of the profit as positive.

Consider a discrete random variable V for which p_i is the probability attributed to the event or hypothesis '$V = v_i$'. The mathematical expectation or *expected value* attributed to V is $\mathcal{E}(V) = \sum_i p_i v_i$, where the sum is taken over all the values deemed possible for V. Let c be the number of units of money the person pays in the contract described in Definition 3.1. The person's expected value of profit is $-c + p \times 1$ units, which is positive for all $c < p$ and negative for all $c > p$. So people will enter into contracts in which they perceive the expected profit to be positive but will not enter into contracts in which they perceive it to be negative.

But why does a person seek to maximize their expected value of profit instead of, say, their most likely value of profit? The only relevance of the expected value is found in the idea that the average profit of a person engaging in a long series of contracts converges to the average of the expected profits in those contracts – whether those contracts are identical or not. (A proof of this is given in Appendix A.) It is only the idea that a person considers he or she would eventually return a profit that would make this approach to betting sensible. Therefore, Definition 3.1 is a definition of probability that is based implicitly on the concept of long-run behaviour, possibly in a sequence of non-identical trials (Willink, 2010c).

This conclusion is noteworthy because many advocates of subjective probability do not seem to think of probability as a long-run concept. Instead they give the impression that a probability can be associated with a single event in isolation from

3.1 Probability and sureness

any wider context. Many such authors omit to define probability at all, while the remainder seem to stop at a definition equivalent to Definition 3.1 without asking the question of why people would bet that way. Answering that question leads us to the conclusion that the definition of probability as degree of belief relies on a concept of long-run behaviour, albeit in a series of problems that might not be identical.

So the concept of long-run behaviour is essential if we are to find meaning in a probability p that someone attaches to a hypothesis. The person is implying that if there were an infinite number of hypotheses each of which he or she favours as much as the hypothesis in question then he or she believes that $100p\%$ of these hypotheses would be true. We arrive at the meaning of an uncertainty interval provided with a figure of sureness of 95%. The measurement scientist is saying to the user of the information that 'this interval is one in a series of equally favoured intervals that would be given in a large set of different measurement problems, and 95% of these intervals in this series will enclose the relevant target values'. A long-run success rate of 95% is envisaged, as was described in Section 1.6.

We see that figures of probability imply and describe long-run behaviour. The converse point may also be made. If the appropriate long-run behaviour does not eventuate then the figures are not correct probabilities. Thus, if many independent hypotheses are each attributed a probability of 0.95 then approximately 95% of these hypotheses must be true, else something in the process by which these attributions were made must be deemed incorrect. Moreover, (by the 'weak law of large numbers' described in Appendix A) if probabilities $p_1, \ldots, p_n$ are attributed to a set of n independent hypotheses then, as n increases without bound, the proportion of these hypotheses that are true must be $\lim_{n \to \infty} \sum_{i=1}^{n} p_i/n$, else the process of deciding upon these probabilities must be seen as wrong.

A frequentist definition

Definition 3.1 paints probability as a personal, subjective, thing. However, that definition is equally applicable when all observers will attribute the same probability to the same hypothesis. One such situation is where (i) the hypothesis is that a certain datum will be obtained and (ii) all observers accept the same model of the data-generating process. This means that the concept of probability described in Definition 3.1 is consistent with the concept used in classical statistics, where probability is not seen as a subjective thing but instead:

the probability p of an event is the relative frequency of its occurrence in an infinite series of independent trials.

Definition 3.1 is broader than this frequentist definition. However, it can act as a definition of frequentist probability if certain conditions are placed on the nature of the event or hypothesis H. To turn the central text of Definition 3.1 into a rather unusual definition of frequentist probability, we need only insert the adjective 'appropriate' before the noun 'hypothesis' and rearrange a few words. The result is the following definition.

> **Definition 3.2** The truth or falsehood of an appropriate hypothesis H will be determined. The *probability p* to be attributed to H is the maximum amount a consistent gambler would pay in order to receive 1 unit if H is found to be true.

This definition is consistent with the frequentist understanding given above because if H is a hypothesis that possesses a frequentist probability then a consistent gambler will bet on H in the manner described if and only if this probability is p.

This logic shows that, contrary to what is implied in most statistical texts, the essential difference between the frequentist and subjective views of probability is not found in the definition! Rather, as is now explained more fully, it is found in the type of hypothesis to which attributing a probability is seen as legitimate. This is related to the idea that some hypotheses only admit a concept of sureness that is subjective.

Legitimate hypotheses for probability statements

It is a source of pride for many New Zealanders (including this author) that the Kiwi mountaineer Sir Edmund Hillary and the Nepalese Sherpa Tenzing Norgay are the first people known to have reached the summit of Mount Everest. These men are certainly the first to have reached the top and returned to tell the tale, but there remains the possibility that the British mountaineers George Mallory and Andrew Irvine reached the summit in 1924, more than a quarter of a century before Hillary and Tenzing. Can that possibility be attributed a probability in either a personal or scientific context?

Let H be the hypothesis 'Mallory or Irvine reached the summit of Mt Everest'. An interested party might state:

$$\text{the probability that I attribute to H is } 0.7 \qquad (3.1)$$

or equivalently

$$\text{my probability of H is } 0.7.$$

This is a statement of personal belief. The person is entitled to believe anything at all with any amount of conviction at all, and I am entitled to believe quite the opposite. So, unless I wish to question the person's honesty, I must accept statement

3.1 Probability and sureness 37

(3.1) as true. However, in itself, statement (3.1) has no scientific meaning: in no way can any experiment be carried out to replicate the conditions that acted on the bodies and minds of Mallory and Irvine to show that seven times in ten they would have reached the summit. It is a statement of personal belief based on whatever the speaker has been influenced by, and it can legitimately change with a glass or two of wine.

This idea of personal or subjective probability is not confined to yes–no questions like that involving the first ascent of Mt Everest: it is equally applicable to a question involving the value of a physical quantity, such as Newton's gravitational constant G. Some readers might be prepared to make a statement like

$$\text{the probability that I attribute to the idea that } G > 6.67384 \times 10^{-11} \text{ m}^3 \text{ kg}^{-1} \text{ s}^{-2} \text{ is 0.5} \quad (3.2)$$

because the figure involved is the 2010 CODATA recommended value (i.e. estimate) of G (NIST, 2010). Just as (3.1) is a statement of personal belief about the unknown but fixed achievements of Mallory and Irvine, (3.2) would be a statement of personal belief about the unknown but fixed value of the fundamental constant G. In the same way that we cannot question the strength of belief of the amateur historian of mountaineering, we cannot deny the truth of this statement about the strength of belief of this physicist. But just as (3.1) is a statement with no scientific meaning, so (3.2) has no scientific meaning. The unknown value of the fundamental constant G is fixed, so in no way can any experiment be repeatedly carried out to show that G takes a value greater than 6.67384×10^{-11} m^3 kg^{-1} s^{-2} five times out of ten or with any other frequency.

The foregoing material briefly describes the concept of subjective probability, which is popular among some statisticians. This author is one of many scientists who are not satisfied with that concept as the basis for the reporting of scientific results, as in a measurement problem. A more objective type of probability is called for. In particular, a concept of probability that is potentially testable seems required. This idea relates to Popper's principle that any theory called scientific must be 'falsifiable', which means that if the theory is false then it can conceivably be shown to be false (Popper, 1968; Magee, 1973). In accordance with this principle, I suggest that a method of assigning probabilities to hypotheses does not meet the standards required in science if we cannot conceive of a means, albeit a hypothetical means, by which the proportion of these hypotheses that are true might be shown to differ from the figure implied.

So I contend that the premises on which an evaluation of measurement uncertainty are based must be potentially amenable to falsification. The classical approach to statistics aims to meet this requirement by only attributing probabilities to hypotheses about outcomes of potentially repeatable processes. Typically, these are hypotheses that relate to events in the future, like the result of the next measurement of G, not to events in the present, like the size of the true value of G; nor to events in the past, like whether Mallory or Irvine reached the summit of Mt Everest. Instead, the trust of the classical statistician in the outcomes of events in the past or present is described in terms of 'confidence' (as we shall see in later sections). For example, imagine we knew nothing about G except that our measurement procedure had probability at least 0.75 of giving an estimate less than G. If the estimate obtained is x then we say we are at least 75% confident that $G > x$. This can be understood as a level of belief or trust, but not as a probability. So, although a probability implies a degree of belief, a degree of belief does not imply a probability. The concept of degree of belief is seen to be broader than this frequentist concept of probability. (That is, a hypothesis relevant in Definition 3.1 is not necessarily appropriate in Definition 3.2.)

In this book we adopt the narrower view of probability. We take Definition 3.1 to be a suitable definition of 'degree of belief' or 'level of sureness' – and so we see Definition 3.1 as completing our definition of a 95% uncertainty interval, Definition 1.1. However, we do not see Definition 3.1 as being a scientific definition of probability. Instead, our definition of probability is Definition 3.2, and we require probability statements to be based on models of data generation that, at least hypothetically, might be found to be incorrect by repetition of the procedure. Yet, as described in Chapter 4, we involve non-standard ideas of 'repetition of the procedure', and so we go beyond the techniques of a traditional statistical analysis. Similarly, we introduce the concept of a relevant set or universe of measurement problems, which relates to the idea that recipients of statements of probability should be receiving something meaningful *to them*. While a probability is a limiting ratio of the 'number of successes' to the 'number of opportunities', this set or universe of opportunities can appear different to different observers.

This section concludes with a brief alternative expression of my view of 'probability' in science. Any conclusion based on probability is based on a probability model, which is a self-consistent assignment of probabilities to a set of events, such as the assignment of the probabilities $\{p_i\}$ to the set of events $\{V = v_i\}$ above. A model must be a model of some thing, an approximation to some reality. A probability model invoked by a classical statistician is analogous to a frequency distribution claimed to approximate the distribution of potential outcomes in repetition of some experiment, and in principle this claim can be tested. This frequency-based model differs from models like those described in the statements (3.1) and (3.2), which are models of nothing but the state of mind of the speaker, which in turn has little relevance to the scientific question.

In accordance with these remarks, the concept of probability adopted in this book obeys a frequency interpretation, albeit an interpretation that does not require the usual narrow understanding of repetition. Other views of probability are not discussed further until Chapter 12.

3.2 Notation and terminology

If the ideas of statistics are to be communicated clearly then several types of mathematical entity must be distinguished. This requires careful use of notation. One dichotomy exists between unknown values that govern possible observations, such as the θ_i quantities in (1.1), and values that exist only after an observation has been made, such as an estimate x_i and the corresponding (unknown) error e_i. In this book these two types of values are indicated using lower-case Greek letters and lower-case Latin letters respectively. A second distinction is to be drawn between a value that actually exists, though it might be unknown, and the concept of a value that is yet to be created through some non-deterministic process. A value that exists is indicated using a lower-case letter, e.g. θ, x or e, while the concept of a potential outcome of some process to take place is indicated using an upper-case letter, e.g. Θ, X or E. For example suppose we are to estimate an unknown fixed value θ_1. The capital letter X_1 represents the concept of the future estimate of θ_1 and the lower-case letter x_1 will denote the actual numerical value taken by X_1. Similarly, suppose making a measurement involves setting the value of an environmental quantity like temperature to be as close as possible to a known figure, x_2, that will be taken as the estimate of this quantity. At the beginning of the process the symbol Θ_2 represents the potential value of the environmental quantity, and at the end of the process the symbol θ_2 represents the unknown actual value of the quantity.

Consider the first of these examples, the measurement of an unknown fixed value θ_1. In the approach advocated in this book, the concept of probability relates exclusively to events that are yet to occur, so the only one of the entities θ_1, X_1 and x_1 that can be made the subject of a probability statement is the potential estimate X_1. A statement like

$$\Pr(X_1 > 0) = 0.8$$

is useful, but statements like $\Pr(\theta_1 > 0.34) = 0.2$ and $\Pr(x_1 > \theta_1) = 0.5$ have no meaning. That is, X_1 is the only one of the three entities that is a *random variable*; a simple, functional, definition of a random variable being 'anything about which we can make a probability statement' or equivalently 'anything given a probability distribution'.

Similarly, consider the second example, where we set an environmental quantity as close as possible to a fixed estimate x_2. The only one of the entities x_2, Θ_2 and θ_2 that can be made the subject of a probability statement is the potential quantity value Θ_2. Hence, Θ_2 is the only one of these entities that is a random variable. On commencing the measurement, we can write

$$\Pr(\Theta_2 > x_2) = 0.5$$

to indicate that there is 50% probability of the environmental value falling above the desired level x_2. But we cannot subsequently write $\Pr(\theta_2 > x_2) = 0.5$ because both θ_2 and x_2 are fixed values, even though θ_2 will be unknown.

This idea of distinguishing between random and non-random variables is very important. Consequently, not only will random variables be denoted using capital letters but probability statements will be written so that the event is expressed as 'subject verb object' as in the basic English sentence. Thus, instead of writing $\Pr(\theta > X) = 0.5$ we will write

$$\Pr(X < \theta) = 0.5$$

to emphasize that X, not θ, is the random entity. Similarly, instead of writing $\Pr(X - 2\sigma < \theta < X + 2\sigma) \approx 0.95$ to indicate that the random interval with limits $X \pm 2\sigma$ is an approximate 95% confidence interval for θ we will write

$$\Pr\left([X - 2\sigma, X + 2\sigma] \ni \theta\right) \approx 0.95,$$

which helps identify the interval as the random entity. (See Section 3.4 for the definition of a *confidence interval*.) One important benefit arising from this discipline will be a clearer demarcation between classical statistical ideas, which are espoused in this book, and Bayesian ideas, which are discussed more fully in Chapter 12.

There is no apology offered for the rigour taken over notation in this book. In this author's opinion, applied scientists will not acquire a shared understanding in matters of statistical data analysis until greater attention is paid to matters of notation. Just as applied scientists can reasonably expect statisticians to take steps outside of their academic culture when considering measurement uncertainty, so physicists and chemists should also be prepared to change their ways where beneficial.

Nor is any apology offered for the fact that statistical notation is sometimes used even though it might be inconsistent with notation more familiar to the reader. This does not mean that applied scientists are expected to abandon symbols such as m and T for an unknown mass and unknown temperature in favour of lower-case Greek letters. Rather, it is suggested that the familiar symbols be annotated to allow the three types of entity to be distinguished. The usual symbol could be retained for an unknown value and the caret ˆ applied to indicate a numerical *estimate* of this value. Also a check ˇ could be applied to indicate the random variable giving rise to this estimate (Willink, 2007a), which is a random variable called the *estimator*. So we might write m for an unknown mass, $\hat{m}$ to indicate the actual estimate of this mass and $\check{m}$ to indicate the estimator of m. The unchecked quantity is called

3.2 Notation and terminology

Table 3.1. *Notation applicable in an experiment involving an unknown parameter that is a mass, a temperature or a resistivity*

	Statistical	Alternative
An unknown fixed value (a number)	θ_i	m, T, ρ
Estimate (a known number)	x_i	$\hat{m}, \hat{T}, \hat{\rho}$
Estimator (a random variable)	X_i	$\check{\hat{m}}, \check{\hat{T}}, \check{\hat{\rho}}$

Table 3.2. *Notation applicable with uncontrolled environmental quantities such as temperature and pressure*

	Statistical	Alternative
Potential quantity-value (a random variable)	Θ_i	$\check{T}, \check{P}$
Environmental quantity-value (an unknown number)	θ_i	T, P
Estimate (a known number)	x_i	$\hat{T}, \hat{P}$

Table 3.3. *Notation applicable with errors*

	Statistical	Alternative
Potential error (a random variable)	E_i	$\check{e}_i$
Actual error (an unknown number)	e_i	e_i

the *outcome* or *realization* of the random variable, so the estimate $\hat{m}$ is the realization of the estimator $\check{\hat{m}}$. Table 3.1 shows this notation and language for mass m, temperature T and resistivity ρ alongside our statistical notation.

The use of the check symbol ˇ to indicate the random variable giving rise to an outcome is appropriate whether or not this outcome acts as an estimate of something. Imagine that the temperature T at which an experiment is carried out is not controlled. Then $\check{T}$ would be the random variable for the potential value of temperature in such a measurement. Table 3.2 shows the notation for temperature and pressure as uncontrolled environmental variables, while Table 3.3 shows the notation for a general component of error.

In this book, analysis within the main body of the text will be carried out using the statistical notation but analysis in the inset examples will generally be described using the alternative notation. The caret ˆ will be used frequently and the check ˇ occasionally. The use of them together, as in $\check{\hat{m}}$, will be kept to a minimum.

Variate and random variable

The fact that the word 'random' has two slightly different meanings was mentioned in Section 2.4. In the mind of the statistician, this word is linked primarily to the idea of a *future* outcome of a process whose output cannot be predicted with certainty. Yet many other scientists will use the terms like 'random data', 'random error' and 'random digits' to describe objects that *have already been* generated by such a process. So it would be easy for the reader to misunderstand the statistical term 'random variable'. For this reason we will follow a suggestion that the term 'variate' be used instead (Marriott, 1990). So we now refer to an entity like X or Θ_i as a *variate*, not as a random variable. (However, an interval variate like $[X - 3\sigma, X - 3\sigma]$ will still be called a random interval.)

Almost every variate to be encountered in this book will be numerical and continuous, which means the outcome of such a variate might be any point in some region of the appropriately dimensioned space, e.g. the real line for a scalar variate. The probability density function of a univariate continuous scalar variate, say Y, will generally be written $f_Y(z)$, the subscript identifying the variate and the argument z being a dummy variable. The letter z is preferred to y because y will often denote the particular numerical value taken by Y. The distribution function of Y is

$$F_Y(w) \equiv \int_{-\infty}^{w} f_Y(z)\, dz,$$

the expected value or mean of Y is

$$\mathcal{E}(Y) \equiv \int_{-\infty}^{\infty} f_Y(z)\, z\, dz$$

and the variance of Y is

$$\mathrm{var}(Y) \equiv \int_{-\infty}^{\infty} f_Y(z)\{z - \mathcal{E}(Y)\}^2\, dz,$$

which will often be denoted $\sigma^2(Y)$.

The two most important distributions in our context are the normal and uniform distributions. If Y has the normal distribution with mean μ and variance σ^2, which is a distribution denoted $\mathrm{N}(\mu, \sigma^2)$, then

$$f_Y(z) = \frac{1}{\sqrt{2\pi}\sigma} \exp\left\{\frac{(z-\mu)^2}{2\sigma^2}\right\}, \qquad -\infty < z < \infty,$$

and if Y has the continuous uniform distribution with lower limit a and upper limit b, which is a distribution that we shall denote $\mathrm{U}(a, b)$, then

$$f_Y(z) = \begin{cases} (b-a)^{-1}, & a < z < b, \\ 0, & \text{otherwise.} \end{cases}$$

3.3 Statistical models and probability models

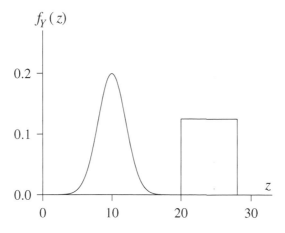

Figure 3.1 Probability density functions of the normal distribution with mean $\mu = 10$ and standard deviation $\sigma = 2$, and the continuous uniform distribution with limits $a = 20$ and $b = 28$.

Figure 3.1 shows examples of these probability density functions.

3.3 Statistical models and probability models

The statistical model underlying any analysis should be explicitly stated. The archetypal model in the measurement of a quantity θ is one of normal error with mean 0 and variance, σ^2. This is written as

$$X = \theta + E, \qquad E \sim N(0, \sigma^2) \qquad (3.3)$$

in which E is the variate for the measurement error e and X is the variate for the measurement estimate x. The statement '$E \sim N(0, \sigma^2)$' is read as 'E has the normal distribution with mean 0 and variance σ^2'. During the measurement, the variates E and X take their numerical values e and x, and we can then write

$$x = \theta + e, \qquad e \leftarrow N(0, \sigma^2), \qquad (3.4)$$

with the statement '$e \leftarrow N(0, \sigma^2)$' being read as '$e$ was drawn from the normal distribution with mean 0 and variance σ^2'. Equations (3.3) and (3.4) are alternative descriptions of the same model, the first involving variates and the second involving the realizations of these variates, the first being familiar to a statistician and the second perhaps reflecting better the understanding of an applied scientist. The first notation might be called *premeasurement notation* because it describes variates with the potential to take values. The second notation might be called *postmeasurement notation* because it describes values that have been generated in the measurement. We will use whichever type of notation is best suited to our context, but we will have more cause to favour postmeasurement notation, especially in Part II.

A model like (3.3) or (3.4) can be applicable even when there are no data. For example, this situation will arise when we refer to the genesis of a fixed error, as in Chapter 4. Many readers might regard such a situation as being outside the realm of 'statistics', and might then prefer the term 'probability model' to 'statistical model'. This author takes a broader view of statistics.

There are many situations where the form of a distribution need not be specified and where only knowledge of the mean and variance is required. So we will let $D(\mu, \sigma^2)$ indicate a distribution with mean μ and variance σ^2. Therefore, the statement '$Y \sim D(0, \sigma^2)$' is equivalent to the statement '$\mathcal{E}(Y) = 0$ and $\text{var}(Y) = \sigma^2$', while the statement '$y \leftarrow D(0, \sigma^2)$' means that the numerical figure y was drawn from some distribution with mean 0 and variance σ^2.

The use of the symbol $\sim$ to mean 'has the distribution' is standard. In contrast, our usage of the symbol $\leftarrow$ is novel. There appears to be no symbol in statistics meaning 'was drawn from the distribution'. Presumably that is because most statistical models are written in terms of variates, as in the premeasurement notation of (3.3).

An alternative means of differentiating between the two ideas is to write that E has the distribution $N(0, \sigma^2)$ but that e has the *parent* distribution $N(0, \sigma^2)$. Similarly, we could write that E has variance σ^2 but that e has *parent* variance σ^2. This language can also be found elsewhere (e.g. Bevington (1969), Coleman and Steele (1999)). We will find it helpful.

Example 1.2 (continued) Recall that an unknown mass was measured using a linear device with unstable parameters. This measurement involved observing the response of the device to reference masses m_1 and m_2 with stated values $\hat{m}_1$ and $\hat{m}_2$. Suppose that the errors $\hat{m}_1 - m_1$ and $\hat{m}_2 - m_2$ can be understood as the outcomes of variates (which is an idea in keeping with the material to be discussed in Chapter 4). Also, suppose that the errors $\hat{m}_1 - m_1$ and $\hat{m}_2 - m_2$ include a shared component with parent variance σ_a^2 and that there are additional individual components of error with parent variances σ_b^2 and σ_c^2 respectively. Last, imagine that the means of the parent distributions of the components of error are zero. Then the relevant statistical model for the estimation of m_1 and m_2 can be written in postmeasurement notation as

$$\hat{m}_1 - m_1 = e_a + e_b,$$
$$\hat{m}_2 - m_2 = e_a + e_c,$$
$$e_a \leftarrow D(0, \sigma_a^2),$$
$$e_b \leftarrow D(0, \sigma_b^2),$$
$$e_c \leftarrow D(0, \sigma_c^2),$$
$$e_a, e_b, e_c \quad \text{were drawn independently,}$$

3.4 Inference and confidence 45

or in premeasurement notation as

$$\check{m}_1 - m_1 = E_a + E_b,$$
$$\check{m}_2 - m_2 = E_a + E_c,$$
$$E_a \sim D(0, \sigma_a^2),$$
$$E_b \sim D(0, \sigma_b^2),$$
$$E_c \sim D(0, \sigma_c^2),$$
$$E_a, E_b, E_c \quad \text{are independent.}$$

3.4 Inference and confidence

The phrase 'statistical inference' is encountered often, but a definition of statistical inference is hard to find. In my mind, a statistical inference involves (i) making an assertion about an unknown quantity and (ii) associating a specified high level of assurance with the correctness of that assertion. The term 'assurance' seems more appropriate than the term 'sureness' that we used earlier because it conveys a greater sense of objectivity and hints at the idea that the statistician is seeking to encourage a sense of trust in the mind of someone else, a client. The fact that the level of assurance is high means that the assertion is seen as a conclusion.

The subject of a statistical inference is generally an unknown fixed parameter like θ. As already argued, a fixed quantity is not a legitimate subject of a probability statement – so the level of assurance is not a direct probability and a different concept of assurance is required. This section describes the concept of assurance that classical statisticians refer to, somewhat unhelpfully, using the word *confidence*.[1]

A simple confidence interval – terminology and success rate

Suppose we are to measure a quantity θ using a procedure for which the error variate E has a probability distribution not depending on θ. Let $e_{0.025}$ and $e_{0.975}$ denote the 0.025 and 0.975 quantiles of this distribution, i.e. the values of z for which $\Pr(E \leq z) = 0.025$ and $\Pr(E \leq z) = 0.975$. The measurement result x is the realization of the variate $X = \theta + E$, so at the beginning of the measurement there is probability 0.95 of obtaining a result lying between the unknown values $\theta + e_{0.025}$ and $\theta + e_{0.975}$, i.e.

$$\Pr(X \in [\theta + e_{0.025}, \ \theta + e_{0.975}]) = 0.95.$$

(In almost all cases $e_{0.025}$ will be negative.) It follows that

$$\Pr([X - e_{0.975}, \ X - e_{0.025}] \ni \theta) = 0.95,$$

---

[1] It is ironic that the word 'confident', which in common English describes something personal, is used in classical statistics to mean something objective. Yet the word 'credible', which means '(objectively) capable of being believed' is used in Bayesian statistics to mean something subjective!

i.e. the random interval $[X - e_{0.975}, X - e_{0.025}]$ has probability 0.95 of enclosing θ. So if $e_{0.025}$ and $e_{0.975}$ are known then we can easily put in place a procedure that will be successful on 95% of occasions: on each occasion we simply quote the numerical interval $[x - e_{0.975}, x - e_{0.025}]$, where x is the relevant observation. If E is symmetric with mean 0 then the limits of this interval are $x \pm e_{0.975}$ and if E is normal with mean 0 and variance σ^2 then they are the familiar limits $x \pm 1.96\sigma$.

The previous paragraph describes the basic idea behind the term '95% confidence interval'. But which interval is, in fact, the confidence interval? Is it the random entity $[X - e_{0.975}, X - e_{0.025}]$ or is it the numerical result $[x - e_{0.975}, x - e_{0.025}]$? The literature seems to be in conflict on this point – which is very unhelpful indeed. Statisticians are likely to express the idea above using the variate X instead of the number x. They might write:

The random interval $[X - e_{0.975}, X - e_{0.025}]$ has probability 0.95 of covering θ whatever the value of θ. So this random interval is a *95% confidence interval for θ*.

Regrettably, the term 'confidence interval' is often used unqualified when describing the corresponding numerical interval $[x - e_{0.975}, x - e_{0.025}]$. This is a major source of confusion, for the correct idea that the random interval $[X - e_{0.975}, X - e_{0.025}]$ has probability 0.95 of covering θ becomes confused with the contrasting notion that the fixed unknown quantity θ has probability 0.95 of lying in the known interval $[x - e_{0.975}, x - e_{0.025}]$. That notion is meaningless in classical statistics because θ is not regarded as possessing a probability distribution.

The use of the term 'confidence interval' to describe the numerical interval obtained in the analysis is widespread, so it does not seem wise to depart far from it. Therefore, in this book we call the numerical interval a 'realized confidence interval'. Thus, the actual interval reported, $[x - e_{0.975}, x - e_{0.025}]$, is called here the *realized 95% confidence interval for θ*. A suitable alternative to the word 'realized' would be the word 'evaluated'.

In this book a lot of emphasis is placed on the use of clear language and notation. This is because the author is convinced that much confusion about statistical concepts has arisen from the poor choice and lazy usage of terms and symbols. So we stress again that we take a confidence interval for θ to be a random interval, not a numerical interval. It is an estimator of θ, not an estimate of θ: the estimate of θ is the realized confidence interval.

Consider the practical value of a confidence interval or confidence-interval procedure[2] for θ. If any procedure has success rate at least 95% whatever the value

---

[2] I find it helpful to think of a confidence interval as a 'procedure'. Imagine casting a net over an invisible target in such a way that you know you will be successful 95% of the time. This procedure is the confidence interval for the location of the target. The area covered by the net after it has been cast is the realized confidence interval.

3.4 Inference and confidence

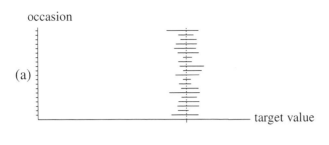
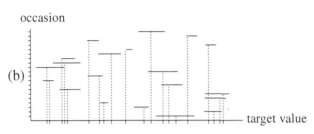

Figure 3.2 Realizations or outcomes of confidence intervals: (a) realizations with a fixed target value; (b) realizations in problems with different target values.

of θ then the success rate must also be at least 95% when this procedure is applied in any series of problems involving different values of θ. This phenomenon is the practical outworking of our concept of probability. Similarly, the idea that the success rate in a relevant series of problems is at least the stated value, 95%, is the merit of the classical approach to *interval estimation* of a parameter, i.e. estimation of the parameter using an interval.

Figure 3.2 illustrates these ideas. Figure 3.2(a) shows many realized 95% confidence intervals for a single target value, while Figure 3.2(b) shows individual realized 95% confidence intervals for many target values. In practice, only one interval is realized in any problem – so the practical merit of putting in place a procedure with probability 0.95 of success in every situation is a success rate of 95% in a series of problems, as in Figure 3.2(b).

Confidence intervals – definitions and concepts

Let us now make our discussion of confidence intervals more general. Consider embarking on the task of measuring some fixed quantity θ when there are other fixed unknowns. The measurement process involves an element of randomness for which we have an adequate statistical model. So it is possible to make a probability statement about a random interval that will extend either side of the potential result of the measurement. We arrive at a non-technical definition of a confidence interval, which is written for our customary level of assurance, 0.95.

Definition 3.3 A *95% confidence interval for θ* is a random interval that has probability 95% of covering θ no matter what the values of all the unknown parameters.

The interval is random because at least one of its limits is random, i.e. is a variate. The requirement that the probability is 95% for all possible values of the unknown parameters is important because meeting this requirement guarantees a success rate of 95% when the procedure is applied in a set of independent problems. If the probability were lower for some values of the unknown parameters then, from the classical point of view, we could not claim to have 95% confidence in the interval realized.

The first duty of the statistician is to avoid making unjustified claims for the procedure employed. The procedure should not perform worse than is implied by the attached level of confidence – but there is no harm done if the procedure performs better than is advertised. For this reason we will describe a confidence-interval procedure with probability of success equal to or exceeding the stated value as a *valid* procedure.

Definition 3.4 A *valid 95% confidence interval for θ* is a random interval that has probability at least 95% of covering θ no matter what the values of all the unknown parameters.

This concept of 'validity' seems important. In fact, some statisticians might remove the word 'valid' from Definition 3.4 and use the result as a definition of a confidence interval per se (e.g. Fisz (1963), Larson (1982)). However, this concept does not seem to be greatly emphasized. Instead we might read of a confidence interval being *exact*, which means that the probability it will contain θ is exactly the specified level, or *conservative*, which means the probability is greater than the specified level, or *anti-conservative*, which means the probability is less than the specified level. We see that a valid interval is one that is exact or conservative for all sets of unknown parameters.

Example 3.1 Consider taking a random sample of predetermined size n from a normal distribution with unknown mean θ and unknown variance σ^2. Let $\bar{Y}$ be the variate for the sample mean, let S be the variate for the usual sample standard deviation and let $t_{n-1,0.975}$ be the 0.975 quantile of the t distribution with $n-1$ degrees of freedom. The random interval $[\bar{Y} - t_{n-1,0.975} S/\sqrt{n},\ \bar{Y} + t_{n-1,0.975} S/\sqrt{n}]$ has probability exactly 95% of enclosing θ no matter what the values of θ and σ^2. So this random interval is an exact 95% confidence interval for θ. If $\bar{y}$ and s are the realizations of $\bar{Y}$ and S then the numerical interval $[\bar{y} - t_{n-1,0.975} s/\sqrt{n},\ \bar{y} + t_{n-1,0.975} s/\sqrt{n}]$ is the corresponding realized exact 95% confidence interval for θ.

3.4 Inference and confidence

Now suppose that the random sample is taken from a normal distribution with mean $\theta + e_{\text{fix}}$, where e_{fix} is a fixed error known only to have magnitude less than or equal to some figure a. Unless $e_{\text{fix}} = 0$, the same random interval has probability less than exactly 95% of enclosing θ, in which case it is an anti-conservative 95% confidence interval for θ (and so it is not valid). However, the longer random interval $[\bar{Y} - t_{n-1,0.975}S/\sqrt{n} - a, \ \bar{Y} + t_{n-1,0.975}S/\sqrt{n} + a]$ has probability at least 95% of enclosing θ. The probability is only equal to 95% in the worst-case scenario where $|e_{\text{fix}}| = a$ so, in practice, this longer random interval is a conservative 95% confidence interval for θ.

The 'confidence level' associated with a confidence interval is the minimum probability that the procedure would generate an interval containing θ, this minimum being taken with respect to all possible sets of values for the unknown parameters. This figure is also called the *confidence coefficient*. We see that a valid 95% confidence interval has confidence coefficient 0.95 or more.

In many situations, the analysis necessary to produce a valid interval is too difficult to be carried out exactly, and we rely on steps that are approximate. If, according to the model, the confidence coefficient is approximately the stated value then the confidence interval is described as *approximate*. Reality dictates that our models describing the generation of our data are only approximations to the actual processes at work. So any claim of exactness in practice would seem unjustified anyway.

Example 3.2 Consider taking random samples of predetermined sizes n_1 and n_2 from normal distributions with means θ_1 and θ_2. For the sample of size n_i, let $\bar{Y}_i$ and S_i be the variates for the sample mean and the usual sample standard deviation. Define the variates

$$S' = \sqrt{S_1^2/n_1 + S_2^2/n_2}$$

and

$$\check{\nu} = \frac{(S_1^2/n_1 + S_2^2/n_2)^2}{(S_1^2/n_1)^2/(n_1 - 1) + (S_2^2/n_2)^2/(n_2 - 1)}.$$

The random interval

$$[\bar{Y}_1 + \bar{Y}_2 - t_{\check{\nu},0.975}S', \ \bar{Y}_1 + \bar{Y}_2 + t_{\check{\nu},0.975}S']$$

has probability approximately 95% of enclosing $\theta_1 + \theta_2$. So this random interval is an approximate 95% confidence interval for $\theta_1 + \theta_2$. (The realization of the variate $\check{\nu}$ is an 'effective number of degrees of freedom'. See Section 6.4.)

After ensuring that the procedure is valid or at least approximate, the statistician should attempt to optimize other aspects of the performance of the estimation procedure. So attention will then focus on the length of the interval. The statistician will endeavour to provide the client with an interval that is short, this interval being more informative about θ than one that is long. The shortness of the intervals generated by a procedure is associated with a statistical concept of *efficiency* – a fully efficient procedure being one that makes full use of the information in the data and the model.

The two aspects of performance – high success rate and small interval length – are in competition. For example we could obtain short intervals by, wrongly, accepting invalid procedures. Therefore, our technique of analysis should be chosen to make the probability of success as little above 95% as possible. This will make the resulting intervals short while keeping the procedure valid.

A question

It is usual to accept a procedure with confidence coefficient reaching a level close to the nominal figure, so a confidence interval constructed in practice might be approximate. But if we know that our methods are only approximate and if our primary responsibility is to generate intervals that are successful on at least 95% of occasions then, to guard against possible detrimental affects of our approximations, should we not employ only procedures with theoretical success rates actually exceeding 95%? This would help prevent the approximate nature of our analyses from invalidating the intervals obtained. Or can we rely on the fluctuations in success rate caused by our approximations to cancel out over the universe of problems relevant to our client? If that were the case then accepting only procedures deemed suitable *after* taking the fact of approximation into account would increase the lengths of the intervals unnecessarily.

This issue is related to the question of what a client can legitimately expect from a procedure that generates uncertainty intervals stated with a specified level of assurance. The author's view is that (a) we should aim to develop procedures that are successful on at least 95% of occasions when applied in practice but that (b) if the actual rate (which can be investigated by simulation under the model and under realistic departures from the model) is slightly lower that 95% for some sets of unknown parameters then the procedure is acceptable.

Premeasurement and postmeasurement assurance

Consider the use of a 95% confidence interval in a measurement problem. If our statistical model is accurate then before the measurement we can be 95% sure that the interval to be generated will contain θ. Does this mean that after the measurement we can be 95% sure that θ lies in the realized interval? In the great majority

3.4 Inference and confidence

of cases, the answer is 'yes'. Unless there is some reason to think otherwise, the amount of faith we can place after the measurement in the idea that we *have been* successful is the same as the amount of faith we placed before the measurement in the idea that we were *going to be* successful. So, unless there is some piece of information that is not fully taken into account in the analysis, we can be 95% sure that θ is contained in the realized 95% confidence interval. This is an important idea. Unless there is some unused piece of relevant information, our postmeasurement level of assurance of success will be the same as our premeasurement level, which was 95%. The interval can then be called a '95% uncertainty interval for θ'. See Section 1.6, in particular Claims 1.2 and 1.3.

One situation where our pre- and post-measurement (or pre- and post-data) levels of assurance differ is where the realized interval extends into a region disallowed for θ (Gleser, 2002). For example suppose one or both of the calculated limits is negative but θ is known from physical considerations to be positive. The existence of a lower bound on θ will not have featured in the statistical model of the measurement error and so will not have been taken into account in the analysis. This is a situation for which the author knows no intellectually satisfying solution. Chapter 15 discusses many of the relevant issues.

The preference for confidence

The idea of a *confidence interval* might wrongly be perceived as being old-fashioned, as predating and being inferior to the Bayesian idea of a *credible interval* (which – as explained in Chapter 12 – relies on the idea that a complete set of prior probability statements can be specified for each relevant unknown parameter). Quite reasonably, that view could be associated with the idea that postmeasurement assurance is preferable to premeasurement assurance. However, difficulty exists with the idea of quantifying postmeasurement assurance in an acceptable way. Rather than being just a modern issue, this issue seems to have been instrumental in the thinking of Jerzy Neyman, who developed the theory of confidence intervals, and in that of his co-worker Egon Pearson. In the book *Comparative Statistical Inference*, which discusses the philosophies and principles of different approaches to statistics, Barnett (1973, pp. 114, 219) writes

[The early papers of Neyman and Pearson on hypothesis testing] show a persistent concern with the importance of both prior information and costs, but an increasing conviction that these factors will seldom be sufficiently well known for them to form a quantitative basis for statistical analysis.

and

Note how criticism of decision theory may operate at two distinct levels. It is either philosophical in nature, in rejecting utility theory and its subjective nature. Alternatively it may take a purely practical form, in denying the possibility (in all but the most trivial problems) of adequately specifying the action space or of eliciting the appropriate loss structure or prior probabilities. It was precisely this latter concern which directed Neyman, Pearson, Fisher and others away from considering prior probabilities and consequential costs to a sole concern for sample data.

So the acceptance of confidence intervals in the statistical community should not be seen as a primitive development but as a step taken in the presence of other views.

3.5 Two central limit theorems

There are at least two central limit theorems that are relevant to the analysis of measurement data. The first involves the well-known idea that if $Y_1, Y_2, \ldots$ are a set of independent identically distributed variates with mean μ and variance σ^2 then as $n \to \infty$ the distribution of the variate $\sum_{i=1}^{n} Y_i/n$ approaches the normal distribution with mean μ and variance σ^2/n. This description is not particularly helpful because as $n \to \infty$ the variance of the distribution approaches zero, in which case the idea that the distribution approaches normality seems irrelevant. A better description is that as $n \to \infty$ the distribution of the variate

$$\frac{\sum_{i=1}^{n} Y_i/n - \mu}{\sigma/\sqrt{n}}$$

approaches the standard normal distribution $N(0, 1)$. This is an informal statement of the classical central limit theorem due to Lindeberg and Lévy (Hoeffding, 1982; Wolfson, 1985b; Stuart and Ord, 1987). This theorem gives a partial justification for treating the mean of a sizeable number of independent results of repeated measurements as having been drawn from a normal distribution.

The second theorem does not require the variates to be identically distributed. If $Y_1, Y_2, \ldots$ are independent variates and if Y_i has mean μ_i and variance σ_i^2 then (unless a finite number of the variates have variances infinitely larger than those of the other variates) as $n \to \infty$ the distribution of the variate

$$\frac{\sum_{i=1}^{n} Y_i - \sum_{i=1}^{n} \mu_i}{\sqrt{\sum_{i=1}^{n} \sigma_i^2}}$$

approaches the standard normal distribution $N(0, 1)$. This is an informal statement of a central limit theorem associated with Liapunov, Lindeberg and Feller (Hoeffding, 1982; Wolfson, 1985a). In our context, this theorem is more useful than the first because a random error can be regarded as the sum of individual elementary errors that correspond to *different* elemental effects, the origins of which are unimportant. The theorem gives a basis for taking the distribution of a random error as being normal. We will call this second theorem the *central limit theorem for non-identical variables*. Example 3.3, which is to follow shortly, gives an example of the marvellous phenomenon it outlines.

These two theorems describe behaviour as n tends to infinity. Of course, n is never infinite in practice. For the identically distributed case, bounds on the difference between the true distribution and the corresponding normal approximation at finite values of n are provided by the Berry–Esseen inequality (Hoeffding, 1982).

3.6 The Monte Carlo method and process simulation

Consider the distribution of a variate $Y = H(Y_1, \ldots, Y_m)$, where H is a specified function and each Y_i is a variate with a known distribution. The distribution of Y is difficult to derive analytically except in the simplest of situations. The Monte Carlo method is an intuitive numerical means for obtaining an approximation to this distribution.

The method involves the straightforward idea that a single value drawn randomly from the distribution of Y is obtained by applying the function H to individual values drawn randomly from the distributions of $Y_1, \ldots, Y_m$. When many such values of Y are generated and formed into a normalized histogram we obtain a close approximation to the probability density function of Y. This procedure can well be called Monte Carlo *simulation* when the function $H(Y_1, \ldots, Y_m)$ represents a process, not just a mathematical relationship.

Example 3.3 Consider a total error variate $E = \sum_{i=1}^{4} E_i$, where each E_i is independent. Suppose that E_1 has the distribution U(0, 4), E_2 has the standard exponential distribution (which is asymmetric), E_3 has the t distribution with five degrees of freedom and E_4 has the standard Laplace distribution. So

$$f_{E_1}(z) = \begin{cases} \frac{1}{4} & 0 < z < 4, \\ 0 & \text{elsewhere.} \end{cases}$$

$$f_{E_2}(z) = \begin{cases} \exp(-z) & z > 0, \\ 0 & z < 0. \end{cases}$$

$$f_{E_3}(z) \propto \left(1 + \frac{z^2}{5}\right)^{-(5+1)/2} \qquad \infty < z < \infty.$$

$$f_{E_4}(z) = \begin{cases} \frac{1}{2}\exp(-z) & z > 0, \\ \frac{1}{2}\exp(z) & z < 0. \end{cases}$$

Individual elements are independently drawn from each of these four distributions and are summed. This is carried out 10^6 times and a histogram is made of the results.

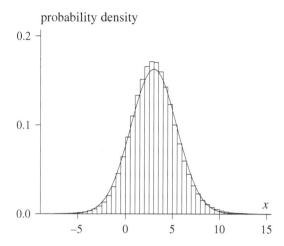

Figure 3.3 A histogram of the distribution of the error variate $E = E_1 + E_2 + E_3 + E_4$, and the normal distribution with the exact mean and variance of E.

The vertical axis is scaled so that the histogram has area 1. The result is shown in Figure 3.3. The histogram can be regarded as providing a close approximation to the probability density function of E. Relevant quantiles of the distribution of E are estimated by the corresponding elements in the ordered set of results.

The means and variances of the E_i variates are:

$$\mathcal{E}(E_1) = 2, \qquad \text{var}(E_1) = \tfrac{4}{3};$$
$$\mathcal{E}(E_2) = 1, \qquad \text{var}(E_2) = 1;$$
$$\mathcal{E}(E_3) = 0, \qquad \text{var}(E_3) = \tfrac{5}{3};$$
$$\mathcal{E}(E_4) = 0, \qquad \text{var}(E_4) = 2,$$

so the mean and variance of E are 3 and 6. The probability density function of the normal distribution with this mean and variance is superimposed on the histogram. The action of the central limit theorem for non-identical variables is evident.

Derivation of the distribution of $Y = H(Y_1, \ldots, Y_m)$ can be interpreted as an act of integration in $m - 1$ dimensions. So the Monte Carlo method as described here can be used as a means of numerical integration.[3] However, the Monte Carlo principle also provides a means of virtual experimentation. Any measurement procedure with known parameters can be simulated if there are adequate statistical models of its non-deterministic steps. This means that we can find a close approximation to the distribution of potential measurement results in the situation of interest, and this

---

[3] The Monte Carlo method described here is a method for evaluating the distribution of a function of random variables. In other contexts, the term 'Monte Carlo method' seems to imply a different objective, which is the evaluation of an integral $\int g(z) f_Y(z)$ over some domain (e.g. Fishman (1996)).

information can be used in the construction of an interval of measurement uncertainty. That approach to the evaluation of uncertainty is described in some detail in Chapter 8.

Monte Carlo methodology can also be used to examine the success rates of various measurement procedures. A measurement can be repeatedly simulated for each set of parameter values in a realistic range, so the confidence coefficient of an estimation procedure can be found (e.g. Hall (2008)). Furthermore, we can simulate many measurements of the same type to examine the success rate over a set of such measurement problems. This is related to the important idea of treating systematic errors 'probabilistically', which is discussed in the next chapter. It is also related to the idea of 'average confidence', which will be discussed in Chapter 5.

4
The randomization of systematic errors

One reason for the perceived failure of classical error analysis in practical measurement problems is the existence of systematic errors. The statistician holding fast to the traditional view of repetition in science cannot deal with an unknown systematic error that is unbounded and can only accommodate a systematic error that is bounded by allowing it worst-case values. For example if the statistical model of the measurement is

$$x = \theta + e_{\text{ran}} + e_{\text{sys}}, \qquad e_{\text{ran}} \leftarrow N(0, 4),$$

where e_{sys} is a systematic error with magnitude known only to be less than 5 units then, under this model, the shortest realized valid 95% confidence interval for θ is

$$[x - 1.96\sqrt{4} - 5, \; x + 1.96\sqrt{4} + 5], \qquad (4.1)$$

which is the interval $[x-8.92, \; x+8.92]$. This approach of using worst-case values seems to be the only honest treatment of systematic errors under the basic rules of classical statistics. However, if our rules of inference somehow permitted us to see e_{sys} as having been drawn from some known probability distribution then we could potentially quote a shorter interval. The benefit would be great in measurements that were dominated by systematic errors. Instead of the length of the interval being approximately proportional to the number of systematic errors, n, the length would be approximately proportional to $\sqrt{n}$.

But in what way must the rules be changed if we are to allow e_{sys} to be interpreted in this way? We have seen that the statement of a meaningful numerical probability requires adherence to some concept of long-run behaviour, whether it be in the hypothetical repetition of a single procedure or in the conduct of a long series of different procedures. So if we are to treat systematic errors as outcomes of variates then a corresponding concept of long-run behaviour must be found. Thus we will now describe a logical basis for the practice of treating systematic errors probabilistically. This involves (i) distinguishing between a 'measurement' and an 'experiment', and then (ii) replacing the concept of success rate in repetition of the same measurement by the concept of success rate over many separate measurements.

4.1 The Working Group of 1980

The problems associated with the existence of systematic errors in measurement led the Bureau International des Poids et Mesures to convene a group of scientists to consider the matter. The text of the report issued by this 'Working Group' provides a clear idea of the group's thinking and intentions (Kaarls, 1980). It describes the basis of an approach that was to be developed for the evaluation of measurement uncertainty. In the abstract of the report we read:

The new approach, which abandons the traditional distinction between "random" and "systematic" uncertainties, recommends instead the direct estimation of quantities which can be considered as valid approximations to the variances and covariances needed in the general law of "error propagation".

And on p. 6 of the report we read:

The traditional distinction between "random" and "systematic" uncertainties (or "errors", as they were often called previously) is purposely avoided here, ...

We see that the concept of error is maintained even though the word 'uncertainty' is preferred.[1]

The Working Group then discusses the basic problem of combining the two types of errors and writes (p. 7):

The only viable solution to this problem, it seems, is to follow the prescription contained in the well-known general law of "error propagation". The essential quantities appearing in this law are the variances (and covariances) of the variables (measurements) involved. This then indicates that, if we look for "useful" measures of uncertainty which can be readily applied to the usual formalism, we have to choose something which can be considered as the best available approximation to the corresponding "standard deviations".

The variances described are variances of measurements, and it is reasonable to interpret these 'measurements' as what we are calling 'measurement results' and 'measurement estimates'. Consequently these variances are the variances of the errors in the measurement results.

---

[1] The question of whether or not the Working Group understood the term 'systematic error' to refer only to a fixed error is not relevant here.

The Working Group then describes three approaches to the specification of the figure of standard uncertainty linked to a systematic error and writes (p. 8):

> In these approaches it is necessary to make (at least implicitly) some assumption about the underlying population. It is left to the personal preference of the experimenter whether this is supposed to be for instance Gaussian or rectangular. Generally speaking, realistic estimates of the limits of uncertainties belonging to a given probability level should be aimed at, and in particular an extremely "conservative" attitude should be avoided.

So, by seeing a systematic component of error as being drawn from some population of errors with a distribution specified by the experimenter, systematic components of error are harmonized with random components. The two types of error are being treated in the same way, so the distinction between them diminishes. This seems to be a major reason for the terms 'random' and 'systematic' now being used less frequently.

The Working Group then provides a formal recommendation that was more succinct than the body of the report. Because of its brevity and formulaic nature, this recommendation does not convey the intention of the Working Group as clearly as the body of the report. So in this chapter we have quoted from the report. Extracts of the recommendation appear in Chapter 14, where we discuss the *Guide* (BIPM et al., 1995), which was written in the light of the recommendation.

Statistical description

Suppose the total numerical error in some measurement result can be written as the sum of random and systematic components,

$$e = e_{ran} + e_{sys}.$$

The report of the Working Group is describing the idea that, just as e_{ran} is seen as the outcome of a variate E_{ran} with some distribution (usually normal), so e_{sys} is seen as the outcome of a variate E_{sys} with a distribution specified by the experimenter. So e is seen as the outcome of a variate

$$E = E_{ran} + E_{sys} \qquad (4.2)$$

whose distribution can be assessed using knowledge of the distributions of its components. Consequently, a realized confidence interval for the target value in the measurement can be obtained.

Example 4.1 The target value is θ and the measurement estimate is the numerical figure x. Suppose that the measurement error $x - \theta$ is expressible as $e_{ran} + e_{sys}$, where e_{ran} is the realization of a normal variate with known variance $\sigma_{ran}^2 = 4$. Suppose, also, that e_{sys} can be deemed to be the outcome of an independent normal variate with variance $\sigma_{sys}^2 = 5$. Let X be the estimator of θ, so X is the variate that gave rise to x. Then, before the measurement, the random interval

$$\left[X - 1.96\sqrt{4+5},\ X - 1.96\sqrt{4+5}\ \right] \tag{4.3}$$

had probability 0.95 of covering θ, and so was a 95% confidence interval for θ. Therefore, the interval $[x - 5.88,\ x - 5.88]$ is a realized 95% confidence interval for θ, and this can be taken as a 95% uncertainty interval for θ.

Interval (4.3) may be usefully compared with the longer interval (4.1) even though the error models in the two measurement problems are not the same.

In this book I (somewhat reluctantly) adopt the approach of the Working Group, which is summarized in (4.2) and exemplified in Example 4.1. The consequences of accepting this approach must be appreciated. The phrase 'had probability 0.95 of covering θ' in Example 4.1 means that the interval evaluated would cover θ on 95% of occasions. So, by adopting this approach, we are implying that there is a relevant set of occasions in Example 4.1 involving normally distributed systematic errors. As is shown in the next section, a logical basis for this can be developed by (a) reexamining the idea of repetition in measurement and (b) involving the idea of a relevant set of measurements. A client who is quoted an interval like (4.3) can then see it as the outcome of a procedure that had probability 95% of generating an interval containing the target value. In the long run, 95% of measurements carried out for the client will be successful. That, surely, is what the client can legitimately expect when receiving '95% uncertainty intervals'.

4.2 From classical repetition to practical success rate

To claim that systematic errors have any relevance at all to the 'uncertainty of measurement' is to imply that the process of measurement includes the activities in which these errors arise. Therefore, the measurement process must be understood to include steps such as (a) the acquisition and calibration of the laboratory equipment, (b) the training of staff and (c) the previous measurement of relevant quantities such as fundamental constants of nature. These may be seen as pre-experimental or 'background' steps in the measurement. These background steps of the measurement contribute systematic components of error, while the subsequent experimental steps give rise to all the random components of error (as well as some systematic components).

The conclusion of the Working Group was that, through the association of variances with systematic errors, the different types of error could be treated in the same way. Therefore, the steps of the measurement process that are associated

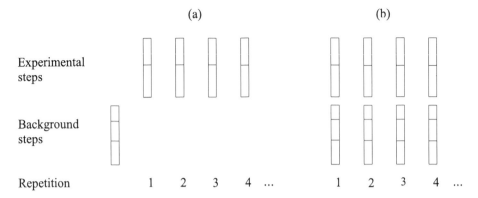

Figure 4.1 Concepts of repetition in measurement: (a) repetition of experimental part of measurement only; (b) repetition of both background and experimental parts of the measurement.

with the different types of errors must also be treated in the same way. This means that the concept of hypothetical repetition of the measurement – which is the concept underlying the idea of a confidence interval – must be extended from the experimental part of the measurement process to include the background part.

These ideas are illustrated in Figure 4.1. Figure 4.1(a) depicts the conventional situation where the only part of the measurement deemed subject to hypothetical repetition is the experimental part. In this case, fixed errors arising in the background steps are not seen as the outcomes of variates: each of these fixed errors becomes an unknown constant. In contrast, Figure 4.1(b) depicts the alternative situation where the concept of repetition applies to the background and experimental parts of the measurement treated as a single whole. Under this view, fixed (and moving) errors can be treated as outcomes of random variables because the ensuing probability statements obey a concept of long-run behaviour.

This important idea can be restated. If our understanding of hypothetical repetition in measurement is illustrated in Figure 4.1(a) then the total fixed error incurred in the background part of the measurement must be seen as an unknown bias to be accommodated using its worst-case values, as in (4.1). However, if our understanding of long-run success rate is based on the idea of repetition that is illustrated in Figure 4.1(b) then both systematic and random errors are seen as outcomes of variates. This is the understanding that permits the analysis carried out in Example 4.1, which led to the interval (4.3).

Activities and stages in the measurement process

So a measurement must now be seen as consisting of background steps as well as experimental steps. Therefore, the generation of error in any result commences

4.2 From classical repetition to practical success rate

Table 4.1. *Acts and stages in the measurement process*

Act	Stage
Adoption of best estimates of fundamental constants	Background
Acquisition of laboratory equipment	Background
Calibration of laboratory equipment	Background
Establishment of protocols	Background
Selection and training of staff	Background
Adoption of reference values	Background
Adoption of external estimates of parameter values	Background
Generation of environmental conditions	Experiment
Estimation of the values of environmental quantities	Experiment
Collection of sample data	Experiment

well before the experimental stages and usually begins in distant laboratories. The acts or actions in which these errors were generated are part of the measurement as a whole. This view is illustrated in Table 4.1, which identifies various actions that generate components of error in a measurement result and suggests the stage in the measurement in which these actions take place.

Let us see how this table relates to Example 1.1, the measurement of the length of the gauge.

Example 1.1 (continued) Recall that the target value is the length l given by

$$l = \frac{l_s(1 + \alpha_s \lambda) + d}{1 + (\alpha_s + \delta)\lambda},$$

where l_s is the length of the standard gauge at 20 °C, α_s and $(\alpha_s + \delta)$ are the coefficients of thermal expansion of the two gauges, λ is the difference in the experimental temperature from 20 °C and d is the difference in lengths of the gauges at the experimental temperature.

The standard gauge exists before the measurement at hand begins. Similarly, the coefficients of thermal expansion of the materials that form the gauges exist outside the measurement. So the unknown values l_s, α_s and δ exist before the measurement. However, the act of adopting estimates of these quantities should be seen as occurring within the measurement. The relevant activity for l_s in Table 4.1 is 'Adoption of reference values' and the relevant activity for α_s and δ is 'Adoption of external estimates of parameter values'. The temperature deviation λ is produced in the experimental part of the measurement in the activity referred to as 'Generation of environmental conditions' and a single estimate of λ is obtained in the activity listed as 'Estimation of the values of environmental quantities'. The contemporary difference in lengths d depends on the temperature and so is created when the environment is created. The estimate of d is obtained when the sample data are collected. These relationships are summarized in Table 4.2.

Table 4.2. *Events in the measurement of the length l*

Event	Period
Creation of l_s, α_s and δ	Before the measurement
Creation of $\hat{l}_s$, $\hat{\alpha}_s$ and $\hat{\delta}$	Background part of the measurement
Creation of λ and d	Experimental part of the measurement
Creation of $\hat{\lambda}$ and $\hat{d}$	Experimental part of the measurement

A practical success rate

The preceding material has taken us some way from the classical statistician's concept of repetition towards the concept of an acceptable success rate in practical measurement problems. The broadening of the definition of 'a measurement' to include background steps means that systematic errors can be treated probabilistically, as was envisaged by the Working Group of 1980. Our analysis can then involve the idea of ensuring a success rate of at least 95% in repetition of a measurement.

Many scientists might argue that the idea of repeated measurement is irrelevant because, in practice, no such repetition takes place: any overall measurement is carried out once and a single result x is obtained. Therefore, it must be asked 'what is the merit of putting in place a procedure that has an adequate success rate under repetition?' This is essentially the same question as 'what is the merit of realizing a valid 95% confidence interval?' The answer was given in Section 3.4: the person receiving intervals calculated by such methods in a long series of differing problems can trust that at least 95% of the intervals will contain the corresponding target values. In the absence of extra, external, information about the suitability of any numerical interval obtained, such an interval seems worthy of the name '95% interval of measurement uncertainty', as implied in Claim 1.3.

The fact that using a procedure with a 95% success rate in every individual situation gives a 95% success rate in a set of different problems will be obvious. However, a proof of this fact will help us make an important point.

Proof Let θ_j be the target value in the jth measurement problem and let $F_j(\cdot)$ be the cumulative distribution function of the corresponding measurement-error variate E_j. The unknown numerical error e_j is seen as being drawn from a population with this distribution function, which means that the unknown quantity $F_j(e_j)$ was drawn from the standard uniform distribution, i.e. $F_j(e_j) \leftarrow U(0, 1)$. So the inequality $0.025 < F_j(e_j) < 0.975$ will hold in 95% of all measurements. This means that the assertion

$$'\theta_j \in [x_j - F_j^{-1}(0.975), \ x_j - F_j^{-1}(0.025)]'$$

will be true in 95% of all measurements. So if, in the jth measurement, x_j is the measurement result and we quote the interval $[x_j - F_j^{-1}(0.975), \ x_j - F_j^{-1}(0.025)]$ then we will achieve a success rate of 95% in a long series of measurements, and this holds even though the measurements in the series might not be identical.

The subscript j identifies a measurement, not a replicate – so this proof makes no reference to repetition of a measurement. We can conclude that, although it is both proper and useful, *the classical statistician's concept of hypothetical repetition of an experiment or measurement is not needed to find meaning in the idea of a success rate achieved in practice*. All that is needed are a definition of 'success' and a definition of the relevant set of occasions or measurements that form the denominator in the success rate or success ratio.

In summary – in this section we have put forward two distinct and complementary ideas that extend the classical statistician's concept of a success rate under hypothetical repeated sampling to a concept of practical success rate in measurement. These are the ideas that (a) the measurement process includes the steps in which systematic errors are generated and (b) the success rate that is relevant in practice is the success rate in some set or some universe of different measurements. Both these ideas allow systematic errors to be seen as arising from parent distributions and as having parent variances. Although it will not be known whether any individual measurement has been successful, the success rate of a procedure can be adequately known if the error models are adequate.

4.3 But what success rate? Whose uncertainty?

The questions in the title of this section relate to a fact that is rarely discussed: *the success rate in measurement problems that involve systematic errors will be different for different parties*. We now explain this idea further and argue that any implied success rate of 95% should be a rate that is relevant to the party being quoted the interval, which is the client, but not necessarily to the party quoting the interval, which is the laboratory. When this idea is acknowledged, the step that we have taken of treating systematic errors as having parent distributions actually becomes easier to justify!

> **Example 4.2** Suppose that a scientist routinely makes many measurements with a certain instrument, and that the dominant component of error on each occasion is an associated fixed error e known only to have magnitude less than 20 units. All other

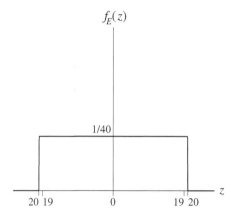

Figure 4.2 Continuous uniform distribution with limits ±20 taken to be the parent distribution for a systematic error, e.

components of error are negligible. A measurement with target value θ is carried out and the numerical result is $x = \theta + e$.

A scientist wanting to treat the fixed error probabilistically might deem it to have been drawn from the uniform distribution with limits ±20, as illustrated in Figure 4.2. If so, he or she would regard the interval $[x - 19, x + 19]$ as a realized 95% confidence interval for θ and would quote this interval as a 95% uncertainty interval for θ. However, the success rate of this procedure as viewed by the scientist is certainly not 95%. If $|e|$ happens to be less than 19 then the success rate will be 100% but if $|e|$ is greater than 19 then the success rate will be zero! To guarantee a success rate of at least 95% from the scientist's point of view he or she must quote the interval $[x - 20, x + 20]$ instead (in at least 95% of the measurement problems).

But this is not so for the scientist's client if the set of measurement problems relevant to that client involves different pieces of equipment, as would be the case if these problems involved approaching different laboratories. The fixed error, in effect, becomes treatable as random because this problem becomes just one of many problems in the client's universe, each of which has systematic errors of its own. The interval quoted can be $[x - 19, x + 19]$. No matter what the value of e, the client can regard this interval as a typical member of a set of intervals arising in relevant measurements, at least 95% of which would be successful. So this shorter interval $[x - 19, x + 19]$ can then legitimately be quoted to the client as a 95% uncertainty interval.

This idea that a success rate of a technique or procedure can be perceived differently by different people seems to be the key to the solution of the Sleeping Beauty paradox, which is described in Appendix B to support this point and for light relief. It is also illustrated in the following example.

Example 4.3 One commuter takes the bus to work when the morning is dry but hires a taxi when it is wet. Over the years he has experienced the bus being on time on 99% of the occasions on which he takes the bus. Another commuter takes the bus when it is wet or windy but cycles to work otherwise. Over the years she has experienced the bus being on time on 80% of the occasions on which she takes the bus, this rate being lower because the bus runs more slowly in wet weather. The day dawns dry and windy, so both commuters will use the bus. They find themselves at the bus-stop discussing how confident they are that the bus will be on time. In the man's universe the bus will be on time with 'probability' 0.99 but in the woman's universe the corresponding figure is 0.8. The two people experience the 'success rate' of the bus company differently because their reference sets of occasions differ.

Let us now return to the measurement conducted by the scientist for his client in Example 4.2. The relevant success rate can be described as the proportion

$$\frac{\text{number of successful measurements in the long run}}{\text{total number of relevant measurements in the long run}}. \quad (4.4)$$

This proportion depends on the identity of the universe of measurements alluded to in the denominator. Is the universe the set of measurements of interest to the scientist, who only uses one particular device? If so, it does not seem legitimate for the scientist to treat the fixed error probabilistically. Or is the universe the set of measurements of interest to the client, with these measurements each involving a different set of errors? If so, the step of 'randomizing' the fixed calibration error in the device becomes more defensible. Evidently, the answer to this question depends on who is making use of the results. This conclusion might seem unsatisfactory, for it implies that a legitimate '95% uncertainty interval' for one observer is not necessarily a legitimate '95% uncertainty interval' for another. So be it! The figure of 95% attached to a statement of measurement uncertainty seems to have no other meaning than that of success rate, and this rate will appear different to different observers.

4.4 Parent distributions for systematic errors

The preceding sections have presented an argument for treating systematic errors in measurement as having been randomly drawn from hypothetical populations of such errors, each of which is described by a parent distribution. As was envisaged by the Working Group of 1980, specifying the variances (and means) of these distributions enables systematic errors to be treated probabilistically through the addition of variances (and means). Subsequent use of the central limit theorem for non-identical variables enables the calculation of an approximate confidence interval. However, if these distributions can be specified more fully then these intervals can be made more accurate. In this section we describe various models for the parent distributions of systematic errors. These are categorized according to whether or not there is a bound on the magnitude of the error.

Bounded distributions

Suppose it can be assumed that the value of a systematic error lies in some known interval. In keeping with this information, any parent distribution used for the systematic error should have zero probability density outside this interval. The distribution or variate can be called bounded. A statistician might describe it as having 'finite support'.

Consider a manufacturer's assertion that the error in the reading of a device has magnitude less that a. Such an error might be supposed to have been drawn from a uniform distribution with limits $\pm a$; see Figure 4.2 and put $a = 20$. A variate with this distribution has probability density function

$$f(z) = \frac{1}{2a}, \qquad |z| < a,$$

and has variance $a^2/3$.

The manufacturer has aimed to make the device as accurate as possible, which means that the error seems more likely to be small than to have a value close to $-a$ or $+a$. In this case the parent distribution might be supposed to have the isosceles triangular form shown in Figure 4.3, where the probability density function is

$$f(z) = \frac{a - |z|}{a^2}, \qquad |z| < a,$$

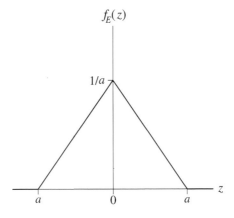

Figure 4.3 Isosceles triangular distribution on $[-a, a]$.

4.4 Parent distributions for systematic errors

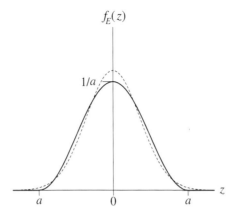

Figure 4.4 Raised cosine distribution on $[-a, a]$ (solid line) and normal distribution with the same mean and same variance (dashed line).

and the variance is $a^2/6$. Alternatively, it might be supposed to be a 'raised cosine distribution' with probability density function

$$f(z) = \frac{1}{2a}\left(1 + \cos\frac{\pi z}{a}\right), \qquad |z| < a,$$

as shown in Figure 4.4, which has variance

$$a^2\left(\frac{1}{3} - \frac{2}{\pi^2}\right).$$

The form of the raised cosine distribution resembles that of the normal distribution except for the fact that it has finite tails. Figure 4.4 also shows the normal distribution with the same mean and variance.

According to the 'principle of maximum entropy' (see Chapter 13), choosing the uniform distribution might be regarded as being minimally committal about the value of the systematic error. However, it seems more important to be maximally conservative about the accuracy of the measurement as a whole, which will often have a total error comprising many components. In that case, the central limit theorem for non-identical variables implies that the most conservative choice would be the distribution with maximal variance, which is the distribution with point masses of 0.5 at $-a$ and $+a$, which has variance a^2. This distribution is seen in Figure 4.5.

Also, a symmetric beta distribution might be used. A symmetric beta distribution that has been scaled and shifted to have support on the interval $[-a, a]$ has probability density function

$$f(z) \propto \left(1 - \frac{z^2}{a^2}\right)^{p-1}, \qquad |z| < a,$$

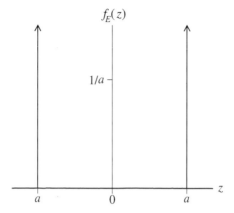

Figure 4.5 Distribution on $[-a, a]$ with maximal variance.

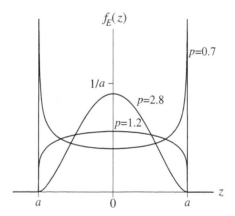

Figure 4.6 Symmetric beta distributions on $[-a, a]$.

mean zero and variance $a^2/(2p + 1)$, with p being a positive shape parameter. Figure 4.6 shows that there is considerable flexibility of shape within this family. The uniform distribution is obtained with $p = 1$ and the 'maximal-variance' distribution is obtained in the limit of $p \to 0$.

Both the raised cosine distribution of Figure 4.4 and the maximal-variance distribution of Figure 4.5 have been put forward as reasonable choices for the distribution of an error that is bounded, as has the uniform distribution. Thus, from one argument we can favour a distribution that is more peaked at the centre than a uniform distribution and from another we can favour a distribution that is much less peaked at the centre than a uniform distribution. The three distributions differ greatly. This highlights the difficulty inherent in proposing parent distributions for

4.4 Parent distributions for systematic errors

systematic errors from such minimal information as a pair of bounds. Many scientists will continue to find the practice of using worst-case values for systematic errors to have more scientific integrity than the idea of manufacturing a distribution from such information.

(It is important to realize that using the maximal-variance distribution for the systematic error is not equivalent to using worst-case values. When using worst-case values, no probability statement is being made about the error at all.)

Unbounded distributions

The most obvious example of an unbounded distribution for modelling a population of systematic errors is the normal distribution. For example, when a device is received from a manufacturer and the calibration error is given as ±2 units it is conceivable that the manufacturer sees the errors in individual devices of that type as being drawn from a distribution that is approximately normal with mean zero and standard deviation 1 or 2 units.

In other situations, the error might be known to have a certain sign. Without loss of generality, we then suppose the error to be positive. A simple unbounded probability distribution that might be used in this case is the exponential distribution with positive parameter λ, which has probability density function

$$f(z) = \lambda \exp(-z\lambda), \qquad z > 0.$$

This probability density function strictly decreases from the value λ at $z = 0$, as shown in Figure 4.7. The mean and variance of this distribution are λ^{-1} and λ^{-2}.

More generally, we might represent the distribution of a positive error variate by a normal distribution truncated below at zero and renormalized. Suppose the mean and variance of the original normal distribution are μ and variance σ^2. It is clear that as $\mu \to \infty$ the truncated distribution approaches the normal

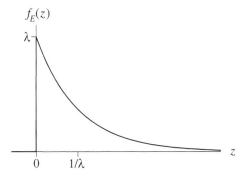

Figure 4.7 Exponential distribution with parameter λ.

distribution $N(\mu, \sigma^2)$. Also, it can be shown that as $\mu \to -\infty$ the truncated distribution approaches the exponential distribution with parameter $\lambda = -\mu/\sigma^2$. (Some examples of truncated normal distributions can be seen in Figures 13.2 and 15.2.)

Do we have choice?

The choice of distribution is not arbitrary. The distribution chosen should either reflect the parent population of systematic errors adequately or lead to a conservative procedure through overestimation of the variance. In this book we work on the basis that an adequate distribution can in fact be proposed. This enables the development of a methodology that is in keeping with the proposal of the Working Group of 1980.

This cautionary point can be restated and emphasized. The parent distribution for an error from some source should provide an adequate approximation to the frequency distribution of errors from that source. If in doubt, the scientist wishing to treat systematic errors probabilistically seems required to use all reasonable competing models and to quote the broadest of the intervals arising under these models. This reflects the idea that it is the first duty of the statistician to provide intervals that are successful at least as often as is implied by the figure 95%. If doubt remains about whether this approach is sufficiently conservative then the only honest approach seems to be to abandon the idea of treating the error probabilistically and to use worst-case values instead.

4.5 Chapter summary

The theme of this chapter is the idea of treating systematic errors as being the outcomes of variates (random variables) with known or supposed distributions. This is one of the most important themes in the book because this step of randomizing systematic errors:

1. is the considered recommendation of the Working Group of 1980, a group of experts set up to address this thorny issue,
2. allows shorter uncertainty intervals to be quoted than would be the case if worst-case values were used,
3. allows uncertainty intervals to be quoted when there are unbounded systematic errors and
4. compels discussion about the practical meaning of the figure 95% associated with a 95% uncertainty interval.

4.5 Chapter summary

The figure 95% must be a success rate of some kind. This step of randomizing systematic errors means that this success rate is taken with respect to a set of many potential problems involving different values for each systematic error. As such, it is the success rate that would be experienced (i) by someone carrying out many measurements with different procedures and different sets of apparatus or (ii) by a person receiving the results from such measurements. This issue of success rate relates to the identity of the set of measurement problems making up the denominator of (4.4), where the quotient must be 95% or more if the intervals are to be called '95% uncertainty intervals'.

5
Beyond the ordinary confidence interval

I have argued that the tools of basic classical statistics are not adequate for the evaluation of measurement uncertainty in practical problems. Special techniques and extensions are required if the principles of measurement are to be properly taken into account. The step taken in Chapter 4 of considering the beginning of the measurement to precede the generation of each error $x_i - \theta_i$ is perhaps the most important of these extensions. This step helps permit the treatment of systematic errors probabilistically, which means that a confidence interval for the target value can be constructed without recourse to the use of worst-case values for systematic errors. In this chapter we also describe two other relevant modifications to the basic idea of a confidence interval. Neither of these ideas seems sufficiently familiar for it to belong in Chapter 3, which addressed foundational ideas in statistics.

Section 5.1 describes a procedure that involves averaging probabilities over the set of possible measurement problems. We call this procedure an *average confidence interval*. It is applicable when the distributions of relevant unknown parameter values in possible measurement problems are adequately known. Section 5.2 describes a procedure that involves a conditional probability statement instead of the unconditional statement that appears, for example, in Definition 3.3. This concept of a *conditional confidence interval* is not novel and might be known to the statistician. Many confidence intervals that are relevant in measurement science seem to be conditional confidence intervals.

5.1 Practical statistics – the idea of average confidence

The valid confidence-interval procedure is the basic tool of the classical statistician for the estimation of a fixed quantity θ by an interval. If the probability that the procedure would be successful is at least 0.95 then infinite repetition of the measurement would result in an interval containing θ on at least 95% of occasions. In practice, an experiment is carried out only once, so the idea of infinite repetition

5.1 Practical statistics – the idea of average confidence

of the same experiment is of little interest in itself. Rather, as has already been described, the practical merit of a confidence-interval procedure resides in the idea of a practical success rate: if the procedure has probability at least 0.95 of being successful with every possible target value then it will be successful in at least 95% of measurements in a long series involving any set of target values. Presumably, that is what a regular client will expect of a procedure purporting to generate 95% uncertainty intervals.

With this idea of a set or universe of measurement problems in mind, we can relax the idea that θ is a fixed unknown quantity isolated from any wider context to the broader idea that θ might be drawn from some frequency distribution. If many or all of the target values making up this distribution are subsequently to be measured then we can conceive of an average amount of confidence over this distribution. Figure 5.1(a) depicts such a distribution in a scenario where manufactured products have actual masses distributed about the nominal figure 1 kg. Figure 5.1(b) shows how this frequency distribution can be seen as the probability distribution of a variate Θ, of which θ is the realization. Thus we can associate a mean level of confidence with the procedure if we know the distribution of all relevant random parameters like Θ. This concept of 'average confidence' represents an extension of classical statistics, and our advocation of it here is one reason why the approach taken in this book might be described as an *extended classical* approach. It is a 'practical frequentist' approach, as opposed to the 'textbook frequentist' approach that was described earlier, where a success rate of at least 95% was to be achieved for all parameter values (Bayarri and Berger, 2004).

For simplicity, our description of this idea of 'average confidence' is given for a type of problem where the target value is the only unknown parameter. The target value in our particular measurement, θ, is now seen as the realization of a variate

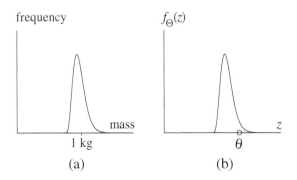

Figure 5.1 (a) A distribution of masses of manufactured products when the nominal mass for the product is 1 kg. (b) The corresponding distribution from which we may see the unknown target value θ as having been drawn. This is the probability distribution of a corresponding variate Θ.

Θ that describes the frequency distribution of target values in some realistic and relevant set of measurements of this type, as in Figure 5.1(b). The success rate of the method in this set of measurement problems will be the success rate averaged over the distribution of Θ. That is,

$$\text{overall success rate} = \int (\text{success rate at } \Theta = z) \, f_\Theta(z) \, dz.$$

So we are led to the following definition of an average confidence interval.

Definition 5.1 A *95% average confidence interval for θ* is a random interval with probability 0.95 of covering θ when the process also involves selecting θ at random from its underlying frequency distribution.

(This can be contrasted with Definition 3.3, which follows the 'textbook' understanding.) We can append the adjectives, 'realized', 'valid' and 'approximate' as appropriate. So loosely speaking,

a realized valid average 95% confidence interval for θ is the result of a procedure that is successful at least 95% of the time in a relevant set of practical problems.

Such an interval satisfies our main requirement for a 95% uncertainty interval for θ.

In some measurement situations we will know enough about the frequency distribution of target values for an average confidence interval for θ to be available, and sometimes this will permit the generation of a shorter interval. As now illustrated, one such situation arises with discretized measurement results.

Example 5.1 Suppose that we measure a quantity θ using a technique that initially incurs normally distributed error E_n with known standard deviation $\sigma = 0.68$ and then involves displaying the result on a device that rounds to the nearest integer. So the final measurement result x is the value taken by the variate

$$X = \lfloor \theta + E_n + 0.5 \rfloor, \qquad E_n \sim N\left(0, 0.68^2\right),$$

where $\lfloor \cdot \rfloor$ means rounding down to the next integer. This situation is illustrated in Figure 5.2 for a particular case where E_n takes the value e_n and where $\theta + e_n$ lies in the interval $[31.5, 32.5]$, so $x = 32$.

The distribution of the measurement error $E = X - \theta$ depends in a periodic fashion on the value of θ. The relevant quantity is the fractional part of θ, which is $\theta - \lfloor \theta \rfloor$. So, for any positive constant a, the probability that the random interval $[X - a, X + a]$ covers θ depends on θ in a similar way. To examine this probability for a possible value of θ we can use the Monte Carlo method to simulate the measurement process.

5.1 Practical statistics – the idea of average confidence

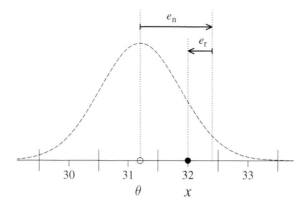

Figure 5.2 Error components when θ is measured with normal error and with a discretizing (rounding) output device. The only known quantity is the estimate $x = 32$. The error $e = x - \theta$ is made up of the underlying normal error e_n and the rounding error e_r, which here are opposite in direction.

The simulation of a single measurement for a possible value of θ involves three steps:

1. Draw $\tilde{e}_n$ from $N(0, 0.68^2)$.
2. Calculate $\tilde{x} = \lfloor \theta + \tilde{e}_n + 0.5 \rfloor$.
3. Determine if the interval $[\tilde{x} - a, \tilde{x} + a]$ contains θ.

The figure $\tilde{e}_n$ is the simulated normal error and the figure $\tilde{x}$ is the simulated measurement result. (In this book, a tilde ˜ is applied to indicate a value arising in a simulation.) These three steps are carried out many times to find the probability that the procedure will generate an interval covering θ using that value of a. We find that setting $a \geq 1.64$ is required to ensure that this probability is at least 0.95 whatever the value of θ. The solid line in Figure 5.3 shows the probability of success as a function of $\theta - \lfloor \theta \rfloor$ when we set $a = 1.64$. A probability of 0.95 is obtained at $\theta - \lfloor \theta \rfloor = 0.36$ and $\theta - \lfloor \theta \rfloor = 0.64$.

However, if we acknowledge that our real requirement is a success rate of at least 95% in a practical set of different problems then another option is open to us. Any long practical set of problems of this type will involve target values with fractional parts spread evenly between 0 and 1 units, else the use of a display with such coarse resolution would seem inappropriate. Let Θ be the variate for the target value that will exist in any one of these problems. Then the variate $\Theta - \lfloor \Theta \rfloor$, which is the fractional part of Θ, has the continuous uniform distribution $U(0, 1)$.

Because of the periodicity, we need only envisage Θ taking random values between 0 and 1 in order to examine the success rate in a corresponding practical set of problems. Any individual trial involves following these steps:

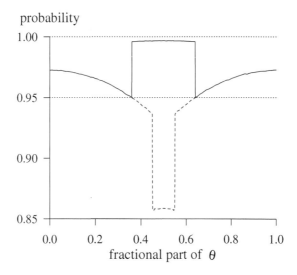

Figure 5.3 Probability of generating an interval containing θ when the process incurs normal error with standard deviation $\sigma = 0.68$ followed by rounding to the nearest integer. Solid line: interval $[X - 1.64, X + 1.64]$; dashed line: interval $[X - 1.45, X + 1.45]$. The fractional part of θ is $\theta - \lfloor \theta \rfloor$.

1. Draw $\tilde{\theta}$ from $U(0, 1)$.
2. Draw $\tilde{e}_n$ from $N(0, 0.68^2)$.
3. Calculate $\tilde{x} = \lfloor \tilde{\theta} + \tilde{e}_n + 0.5 \rfloor$.
4. Record a success if the interval $[\tilde{x} - a, \tilde{x} + a]$ contains $\tilde{\theta}$.

Simulation shows that an overall success rate of 0.95 is obtained by setting $a = 1.45$. The dashed line in Figure 5.3 shows the probability of success with this value of a as a function of $\theta - \lfloor \theta \rfloor$. The average probability over the interval $[0, 1]$ is 0.95. Therefore, the random interval $[X - 1.45, X + 1.45]$ is a 95% average confidence interval for θ and, in the problem at hand, the interval $[x - 1.45, x + 1.45]$ is a realized 95% average confidence interval for θ. The value 1.45 is considerably less than the value 1.64, so – by recognizing that average confidence is the type of confidence required – we have been able to quote a shorter interval than the textbook frequentist statistician.

This analysis has shown that a practical frequentist statistician can treat the target value as having been random, as being the outcome of a variate. Subsequently, what is required is an approximation to the distribution of target values in a set of problems that is realistic and relevant to the person being quoted the uncertainty interval. So, somewhat like the Bayesian statistician of Chapter 12, the practical frequentist statistician is seeking to use an existing distribution underlying θ. However, the legitimate specification of such a distribution seems difficult. What knowledge or information might we possess that could permit it? (Example 5.1, for instance, only involved the weaker claim that we knew the distribution of the *fractional part* of Θ.) A statistician might reluctantly conclude that – in a purely statistical context – the idea of average confidence is only applicable in a small number of situations. This lack of applicability will be one reason why the concept of 'average confidence' is not generally discussed in statistics.

Despite its lack of relevance to the typical statistics problem, the concept of average confidence is useful in measurement problems. This is because many scientists seem prepared to make probability statements relating to various systematic influences on measurement results, and some of these influences will be parameters of the measurement. The values of these parameters might then be seen as outcomes of variates.

Example 1.1 (continued) Recall that the length intended to be measured is given by

$$l = \frac{l_s(1 + \alpha_s \lambda) + d}{1 + (\alpha_s + \delta)\lambda},$$

where δ is the difference between the coefficients of thermal expansion of the two gauges. The quantity δ existed before the background steps of the measurement, so it exists outside of the measurement. It is a therefore a parameter of the measurement. Also, its estimate is the fixed value $\hat{\delta} = 0$. It follows that the error $\hat{\delta} - \delta$ must be seen as an unknown constant in this whole measurement – and not simply as an unknown constant in the experimental part of the measurement. So we cannot apply the first argument given in Section 4.2 for treating this systematic error as an outcome of a variate. We must rely on the second argument, which involved recognising that the practical success rate of interest to a client will be the success rate in a series of distinct measurement problems.

Thus, to obtain a 95% uncertainty interval for l that does not include worst-case values for δ we must include the broadest idea of 'repetition' alluded to in Section 4.2: the figure of 95% is permitted to be the success rate in a set of measurements of the same type, not just in replication of the same measurement. If the error $\hat{\delta} - \delta$ is to be treated as having a parent distribution then we must envisage a realistic real or hypothetical set of measurements of this type, each involving a different value for δ. The error $\hat{\delta} - \delta$ can then be treated as the outcome of a variate. The uncertainty interval calculated would be the realization of a 95% average confidence interval because the success rate of 95% applies with reference to a set of different measurements of this type.

5.2 Conditional confidence intervals

The concept of *conditional confidence* relates to the fact that our level of assurance in any interval can depend on other information obtained during the statistical experiment. This concept is not discussed in most introductory statistics texts, but it seems relevant in measurement problems because an estimate of an environmental variable is information obtained during the measurement. We begin with a definition.

Definition 5.2 A *95% conditional confidence interval for* θ is a random interval that, conditional on the value taken by some observable auxiliary variate, has probability 95% of covering θ no matter what the values of all relevant unknown parameters.

If a 95% conditional confidence interval is available whatever value is taken by the auxiliary variate then the overall success rate must be at least 95%. (Definition 5.2 might not satisfy a statistician, who would choose to define a conditional confidence interval in terms of an 'ancillary statistic', not an auxiliary variate (Kiefer, 1982).)

The key difference between a conditional confidence interval and the ordinary confidence interval of Definition 3.3 is that the probability statement now applies *after* observing the value taken by the auxiliary, extra, variate, i.e. it is a statement of conditional probability. This idea is illustrated in the following highly fictitious example, which first describes an ordinary confidence interval and then shows how observing the value taken by a relevant variate permits evaluation of a conditional confidence interval. The realized conditional confidence interval is shorter and hence more informative.

Example 5.2 A mass m will be measured in a laboratory where the temperature is known to fluctuate between two limits, 20 ± 3 °C. Mass is represented in grams and temperature is represented in degrees Celsius. Let T be the temperature at the time of measurement and let the midpoint of the temperature interval, 20, be the estimate of this temperature. The technique of measuring the mass is sensitive to temperature in such a way that a deviation from 20 introduces an error of $1.8(T-20)$. Let m_{meas} denote the measurement estimate. (The reason we do not use the notation $\hat{m}$ at this point will become clear.) We can write

$$m_{\text{meas}} = m + 1.8(T-20) + e_{\text{other}},$$

where e_{other} is the error from all other sources. The temperature T can be thought of as the realization of a variate $\check{T}$ bounded between 17 and 23. We suppose that $\check{T}$ is attributed the uniform distribution between these limits. We also suppose that e_{other} is drawn from the normal distribution with mean zero and standard deviation 2. That is, e_{other} is the realization of a variate $\check{e}_{\text{other}}$ with the distribution N(0, 4).

In effect, the measurement of m is a process in which the variates $\check{T}$ and $\check{e}_{\text{other}}$ are realized and take the values T and e_{other}. So the measurement estimate m_{meas} is the realization of a variate

$$\check{m}_{\text{meas}} = m + 1.8(\check{T}-20) + \check{e}_{\text{other}},$$

5.2 Conditional confidence intervals

where $\check{T} - 20 \sim U(-3, 3)$ and $\check{e}_{\text{other}} \sim N(0, 4)$. The total error variate is the sum of a uniform variate and an independent normal variate. The distribution of such a variate is described in Appendix C. The 0.025 and 0.975 quantiles of the total error are found to be -5.11 and 5.11.

So the random interval $[\check{m}_{\text{meas}} - 5.11, \check{m}_{\text{meas}} + 5.11]$ has probability 0.95 of enclosing m and is a 95% confidence interval for m. The realization of this interval, which is $[m_{\text{meas}} - 5.11, m_{\text{meas}} + 5.11]$, can be taken as a 95% uncertainty interval for the unknown mass m. This interval has length 10.22 g.

Now suppose we are able to measure the temperature with negligible uncertainty. Then we know that in the measurement $\check{T}$ took the value T. Consequently, we can apply probability statements that hold subject to this knowledge or condition. The appropriate statistical model can be written as

$$\check{m}_{\text{meas}} | (\check{T} = T) = m + 1.8(T - 20) + \check{e}_{\text{other}},$$
$$\check{e}_{\text{other}} \sim N(0, 4).$$

The | mark is to be read as 'on the condition that'. So, conditional on the outcome of the random temperature being known to be T, the point estimator of m is

$$\check{m}_{\text{meas}} + 1.8(T - 20)$$

and the random interval

$$[\check{m}_{\text{meas}} + 1.8(T - 20) - 3.92, \check{m}_{\text{meas}} + 1.8(T - 20) + 3.92]$$

has probability 0.95 of enclosing m. Therefore a realized 95% conditional confidence interval for m is $[\hat{m} - 3.92, \hat{m} + 3.92]$ with $\hat{m} = m_{\text{meas}} + 1.8(T - 20)$, and with T now being known. This interval has length 7.84 g. Not surprisingly, observing the temperature has enabled us to quote a shorter 95% uncertainty interval.

From this example, we see that a conditional confidence interval can be shorter than an unconditional confidence interval. We can also infer that a conditional confidence interval might be available when an unconditional interval is not. For example, imagine that we did not know the limits 20 ± 3 °C for the temperature T or were not prepared to assign the temperature variate $\check{T}$ a uniform distribution between these limits. In that case, we could not calculate the unconditional interval applicable with unknown T, but we could still calculate the conditional interval if we knew T.

The kind of situation described in the second part of Example 5.2 will be familiar to the measurement scientist. Often an environmental quantity like temperature will be measured during the experiment to take into account its influence on a result. The value of this quantity at the relevant time will be the outcome of some non-deterministic environmental process but, because this quantity is measured, no knowledge of this random process will be required. In this way, the idea of conditional confidence described here is fundamental to the use of statistical methods in measurement. The reader may well have been applying this concept for many years without realising that it is not found in most statistics texts.

Example 1.1 (continued) The length to be measured is

$$l = \frac{l_s(1 + \alpha_s \lambda) + d}{1 + (\alpha_s + \delta)\lambda}.$$

The quantity λ is the difference in the temperature at the time of the experiment from 20 °C. The temperature in the laboratory will fluctuate a little, so λ should be seen as the outcome of a variate $\check{\lambda}$. Measuring λ at the appropriate time means that the distribution of $\check{\lambda}$ does not need to be known. The 95% uncertainty interval for l generated will be the realization of a 95% conditional confidence interval for l because the relevant probability statement applies under the condition that $\check{\lambda} = \lambda$.

5.3 Chapter summary

This chapter completes Part I, which has been devoted to a discussion of principles relevant to measurement uncertainty and probability. In this chapter we have described two ideas that are relevant to measurement problems but which are not found in the typical statistics text. The first is the idea of permitting a parameter to be seen as having a known parent distribution applicable in the set of problems over which we require a success rate of 95%. An interval calculated using this principle can be called a realized 95% *average confidence* interval. This idea that some parameter values can be treated as the outcomes of variates is similar to the idea that the values of systematic errors can be treated as the outcomes of variates, which was described in Chapter 4.

Some statisticians might feel uncomfortable with this concept of 'average confidence'. However, it appears to be a legitimate tool for meeting the reasonable expectation of a client in a measurement situation, which must be that at least 95% of the 95% uncertainty intervals received by the client in a relevant set of problems contain the corresponding target values. The idea of average confidence is an extension of the classical idea of confidence to meet this requirement.

The second idea is one of a confidence interval involving a probability statement applicable on the condition that another variate takes a known value. This kind of interval can be called a 95% *conditional confidence* interval. Uncertainty intervals generated in measurement problems where an environmental quantity is measured will often be based on such statements of probability. Thus, many uncertainty intervals calculated in practice will, in fact, be the realizations of conditional confidence intervals.

5.3 Chapter summary

To some extent, the ideas of average confidence and conditional confidence have been described so that the readers can see various differences between the logic featuring in introductory texts on statistics and a logic that seems legitimate and necessary in measurement problems. The heart of the matter is the basic requirement that we place on any procedure for calculating 95% uncertainty intervals. This requirement is that at least 95% of the intervals calculated in a long series of relevant measurements contain the values of the measurands. This reasonable requirement permits the considered use of the concepts described in this chapter. It also admits the concept of randomizing systematic errors, which was discussed in Chapter 4. The use of these concepts is the main reason for viewing the approach taken in this book as being one of extended classical statistics.

Part II

Evaluation of uncertainty

I am ill at these numbers ...
William Shakespeare
Hamlet : Act II, Scene 2

I of dice possess the science
And in numbers thus am skilled.
From the *Mahabharata*

Give me a slate and half an hour's time, and I will produce a wrong answer.
George Bernard Shaw
Cashel Byron's Profession (preface, 1901)

I am ashamed to tell you to how many figures I carried these computations, having no other business at the time.
Isaac Newton

6
Final preparation

Part II of this book describes and discusses methods for the evaluation of measurement uncertainty. This chapter completes our preparation for this task by summarizing the principles put forward in Part I, establishing some basic results, and considering two important steps – the statement of the measurement model and the treatment of normal errors with unknown parent variances.

Section 6.1 presents a list of the most important principles advocated in Part I. Many of these principles form the logical basis of the techniques of uncertainty evaluation to be described. Section 6.2 gives two associated results that show how confidence intervals for the target value can be realized: the first of these results being used in most of the analyses to follow. Section 6.3 describes different ways of stating the measurement model and places emphasis on the use of postmeasurement notation, which seems more relevant to most scientists than premeasurement notation. Section 6.4 addresses the problem of treating normal errors with unknown variances when, as is common in measurement, these errors form only a subset of the components of the total measurement error.

6.1 Restatement of principles

Our summary of the material presented in Part I takes the form of a list of important assertions. The list should be read in order.

1. It is important to see the measurement as having an ideal result, which is the 'target value' θ. A quantity that does not exist cannot feature in a probability statement, so without the existence of a target value there is no logical basis for invoking the statistical idea of a confidence interval.
2. The concept of 'measurement error' is indispensable. This error is $x - \theta$ where x is the measurement result and θ is the target value.

3. Uncertainty and error are not the same thing. The sign and size of the error are unknown. In contrast, the amount of uncertainty must be known – else it cannot be stated. A statement of measurement uncertainty is a statement about the potential size of the error.
4. A measurement is 'successful' if the calculated interval of uncertainty contains the target value.
5. The frequentist definition of probability is consistent with the Bayesian definition of probability as 'degree of belief'. The difference between the two views of probability is found in the types of hypotheses that are regarded as legitimate subjects of probability statements.
6. The definition of probability as 'degree of belief' is based implicitly on a concept of long-run behaviour. The definition relies on the concept of an expected value of profit, and an expected value is only relevant through the idea of long-run behaviour.
7. Consequently, even if a Bayesian approach to inference is adopted, attaching probabilities of 95% to the uncertainty intervals calculated in independent measurements implies that a success rate of 95% is anticipated.
8. The success rate that is relevant to a client might not be the success rate that is relevant to a laboratory because the client and the laboratory might be interested in different sets of problems.
9. Placing the focus on the success rate that is relevant to the client strengthens the logical basis for treating systematic errors as outcomes of variates (random variables). Such a treatment of systematic errors was envisaged by an authoritative group of measurement experts at a time when a digression from the techniques of classical error analysis was sought.
10. A 95% confidence interval for the target value θ is a random interval with probability 0.95 of containing θ no matter what are the unknown values of other fixed quantities in the measurement. This concept relates to the potential outcome of a measurement process containing one or more random influences. After the measurement has been made, the random influences have acted and the randomness has gone. So the statement of probability does not apply to the numerical interval that is the realization of the confidence interval. Instead our assurance or trust in the idea that the numerical interval then contains the target value is described by the word 'confidence'.
11. When 95% confidence intervals are calculated in a real or hypothetical relevant series of measurements, the success rate is 95%. So we can have a level of assurance of 95% in any one of the individual realized intervals provided that there is no additional information that calls into question its individual reliability. If this condition is satisfied, the interval can be regarded as a 95% interval of measurement uncertainty or a '95% uncertainty interval'.

12. The requirement of at least 95% success in a relevant hypothetical set of practical problems does not limit us to intervals that obey the classical concept of confidence just described. Two additional concepts are relevant. First, a parameter might be considered to have a known distribution over a relevant set of problems and, second, many practical measurement problems will involve observing environmental quantities. The appropriate variants of the ordinary confidence interval are the average confidence interval and the conditional confidence interval respectively. Consequently, in the absence of additional information, the realization of a 95% average confidence interval or conditional confidence interval can also be called a '95% uncertainty interval'.

6.2 Two important results

As described in point 11 above, a realized valid 95% confidence interval for the target value θ can be regarded as a 95% uncertainty interval for θ unless there is additional information to call into question its reliability. We now show how a valid 95% confidence interval for θ can be realized when the distribution of the measurement error is known. Our description supplements the derivation of a realized confidence interval given in Part I.

Consider the evaluation of a confidence interval as a means of estimating a parameter θ using our result x. The idea behind the formation of this interval is that we are to include all possible values for θ for which a result as extreme as x might realistically have been observed. So we will include any candidate value θ_c for which the difference $x - \theta_c$ is neither too positive nor too negative. As in Section 3.4, let $e_{0.025}$ and $e_{0.975}$ denote the 0.025 and 0.975 quantiles of the parent distribution of the measurement error $e = x - \theta$. (Recall that $e_{0.025}$ will almost always be negative.) If θ_c were equal to θ then the event $X > \theta_c + e_{0.975}$ would occur on only 2.5% of occasions. Thus, with a degree of trust of 0.975, we can exclude a candidate value θ_c less than $x - e_{0.975}$. Similarly, if θ_c were equal to θ then the event $X < \theta_c + e_{0.025}$ would occur on only 2.5% of occasions. So, with a degree of trust of 0.975, we can exclude a candidate value θ_c greater than $x - e_{0.025}$. Thus, with a degree of trust of 0.95, we can exclude all candidate values outside the interval $[x - e_{0.975}, x - e_{0.025}]$.

To show that this informal kind of thinking has in fact generated a realized 95% confidence interval for θ, we note that $\Pr(X - e_{0.975} < \theta) = 0.975$ and $\Pr(X - e_{0.025} > \theta) = 0.975$, so that the intervals $[X - e_{0.975}, \infty)$ and $(-\infty, X - e_{0.025}]$ are 97.5% confidence intervals for θ. These intervals can be called *right-infinite* and *left-infinite* intervals. A central 95% confidence interval for θ is the intersection of a left-infinite 97.5% confidence interval for θ and a right-infinite 97.5% confidence interval for θ. In our case, the intersection is $[X - e_{0.975}, X - e_{0.025}]$. This random interval has probability 0.95 of covering θ, and its realization is the numerical interval obtained above.

We are then able to state the following result.

Result 6.1 Let x be an estimate of a fixed quantity θ obtained using a technique with a random element, and let $e_{0.025}$ and $e_{0.975}$ be the 0.025 and 0.975 quantiles of the distribution from which the error $e = x - \theta$ was drawn. A realized 95% confidence interval for θ is $[x - e_{0.975}, x - e_{0.025}]$.

This result forms the basis for the methods of uncertainty evaluation described in the following chapters. The interval obtained can be quoted as a 95% uncertainty interval for θ in the absence of other information relating to its reliability. Two familiar special cases of this result will be used in Chapter 7: if the parent distribution of e is normal with mean 0 and variance σ^2 then the 95% interval is $[x - 1.96\sigma, x + 1.96\sigma]$ while if the distribution is symmetric with mean 0 then the interval for θ is $[x - e_{0.975}, x + e_{0.975}]$.

As indicated in Part I, the confidence level of a procedure relates to the long-run success rate of the procedure. So we may also state the following result, which is stronger than Result 6.1 because an interval covering 95% of the parent distribution of e does not have to be found exactly.

Result 6.2 Let x be an estimate of a fixed quantity θ obtained using a technique with a random element, and let $[a, b]$ be the numerical output of a procedure that generates intervals containing a random fraction P of the parent distribution of the error $x - \theta$. If $\mathcal{E}(P) \geq 0.95$ and if the process of generating a and b is independent of the process of generating x then $[x - b, x - a]$ is a realized valid 95% confidence interval for θ.

To show that this result holds, we let A and B denote the variates for a and b, and write $\Pr([X - B, X - A] \ni \theta) = \Pr(A \leq X - \theta \leq B) \geq 0.95$.

Again, provided that the interval obtained is known a posteriori to be as reliable as the typical interval that would be generated by this procedure, the interval can be quoted as a 95% uncertainty interval for θ. We will see in Section 7.4 that this result enables the number of trials required in one type of Monte Carlo analysis to be much smaller than many readers might suppose.

6.3 Writing the measurement model

It is important to develop an explicit mathematical model of the measurement. This model will include the measurand equation $\theta = \mathcal{F}(\theta_1, \ldots, \theta_m)$, the estimate

6.3 Writing the measurement model

equation $x = \mathcal{F}(x_1, \ldots, x_m)$, and appropriate equations describing the generation of the errors $e_1, \ldots, e_m$. Consider the situation where e_i is regarded as being drawn from a distribution with mean zero and variance σ_i^2. The model might be written as

$$\theta = \mathcal{F}(\theta_1, \ldots, \theta_m),$$
$$x = \mathcal{F}(x_1, \ldots, x_m),$$
$$x_1 \leftarrow D(\theta_1, \sigma_1^2)$$
$$\vdots$$
$$x_m \leftarrow D(\theta_m, \sigma_m^2).$$

These equations describe the generation of the data, not the generation of the errors. Our construction of a confidence interval will involve combining the parent distributions of the errors, so a more helpful representation seems to be

$$\left.\begin{array}{rl}\theta &= \mathcal{F}(\theta_1, \ldots, \theta_m), \\ x &= \mathcal{F}(\theta_1 + e_1, \ldots, \theta_m + e_m), \\ e_1 &\leftarrow D(0, \sigma_1^2) \\ &\vdots \\ e_m &\leftarrow D(0, \sigma_m^2). \end{array}\right\} \quad (6.1)$$

A statistician would want to write this in terms of variates, and so would write:

$$\theta = \mathcal{F}(\theta_1, \ldots, \theta_m),$$
$$X = \mathcal{F}(\theta_1 + E_1, \ldots, \theta_m + E_m),$$
$$E_1 \sim D(0, \sigma_1^2)$$
$$\vdots$$
$$E_m \sim D(0, \sigma_m^2),$$

which is (6.1) in premeasurement notation.

It might also be appropriate to define variables in the order in which they arise in the measurement process. The statements describing the generation of the errors would then precede the statement defining the estimator X. Also, we would want the description to explicitly define the measurement error $E = X - \theta$, because it is the distribution of E that is used when forming the confidence interval. Making these changes and reverting to postmeasurement notation gives a description like:

$$e_1 \leftarrow D(0, \sigma_1^2)$$
$$\vdots$$
$$e_m \leftarrow D(0, \sigma_m^2),$$
$$x = \mathcal{F}(\theta_1 + e_1, \ldots, \theta_m + e_m),$$
$$e = x - \mathcal{F}(\theta_1, \ldots, \theta_m).$$

These are four different ways of representing the same model of measurement. Preference is partly a matter of personal choice but, in our context, a model that focuses on the parent distribution of the overall error seems appropriate. Given that $\leftarrow$ is used to mean 'is drawn from', the last description reads like a chronology: in keeping with the sequence of events in the measurement, the overall error is incurred after the contributing errors. So this representation is particularly useful when simulating the measurement.

A preferred style

We have drawn a distinction between the premeasurement notation of the statistician, who would write $X = \theta + E$ with X and E being variates, and the postmeasurement notation of the general scientist, who would write $x = \theta + e$ with x and e being numerical figures, e being unknown. In Part I we favoured neither, but now we prefer the use of postmeasurement notation because this seems more familiar to applied scientists. We will also adopt some corresponding language. The adjective 'parent' will be used freely with postmeasurement quantities. For example the mean and variance of E will be called the parent mean and parent variance of e. Also, we will introduce the subscript 'p' to indicate 'parent'. Thus $\mathcal{E}(E) = \mathcal{E}_p(e)$, $\text{var}(E) = \text{var}_p(e)$ and $\sigma^2(E) = \sigma_p^2(e)$.

These steps might seem unnecessary because language such as 'the variance of e' and notation such as $\text{var}(e)$ or $\sigma^2(e)$ will seem unambiguous to many readers. For example it is clear that in a classical analysis the phrase 'the variance of e' – which has no proper meaning – refers to 'the variance of E, the variate giving rise to e'. However, in a Bayesian analysis the phrase 'the variance of e' would have meaning, and this meaning might differ from that of 'the variance of E'. So there is potential for misunderstanding unless care is taken.

A model example with an example model

It is now appropriate to introduce an example that will be reused several times in the following chapters. At this point, we just formulate the measurement problem and state the model.

6.3 Writing the measurement model

Example 6.1 The distance d between tiny objects at two points on a plane is measured using a device situated at the origin. The distance is given by

$$d = \sqrt{r_1^2 + r_2^2 - 2r_1 r_2 \cos \phi},$$

where r_1 and r_2 are the lengths of the vectors from the origin to the two points and ϕ is the angle between these vectors. See Figure 6.1. Estimating the distance of a point from the origin incurs a fixed relative error with standard deviation 0.2% and a random error with standard deviation 0.1 m. The angle is estimated by averaging $n = 7$ readings each of which has independently incurred normal error with parent mean zero and unknown parent standard deviation σ.

The units of length are metres and the units of angle are radians. The model of the measurement given in a style similar to that of (6.1) is:

$$d = \sqrt{r_1^2 + r_2^2 - 2r_1 r_2 \cos \phi},$$

$$\hat{d} = \sqrt{\hat{r}_1^2 + \hat{r}_2^2 - 2\hat{r}_1 \hat{r}_2 \cos \hat{\phi}},$$

$$\hat{r}_1 - r_1 = r_1 e_a + e_b, \tag{6.2}$$

$$\hat{r}_2 - r_2 = r_2 e_a + e_c, \tag{6.3}$$

$$\hat{\phi} - \phi = e_d,$$

$$e_a \leftarrow D\left(0, 0.002^2\right), \tag{6.4}$$

$$e_b \leftarrow D(0, 0.1^2), \tag{6.5}$$

$$e_c \leftarrow D(0, 0.1^2), \tag{6.6}$$

$$e_d \leftarrow N(0, \sigma^2/7), \tag{6.7}$$

e_a, e_b, e_c, e_d have been drawn independently.

(The errors listed do not have the same dimensions. The dimension of a parent variance is the dimension of the squared error. For example $\text{var}_p(e_b) = 0.01 \text{ m}^2$.)

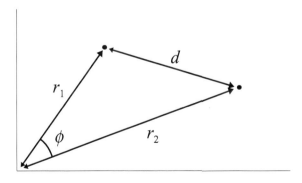

Figure 6.1 The geometry for the measurement of the distance d between two points.

In later chapters, when using this model to evaluate the measurement uncertainty, our attention will focus on (6.4)–(6.6), in which the forms of the distributions are not yet specified. First we consider (6.7), where the underlying variance σ^2 is unknown.

6.4 Treating a normal error with unknown underlying variance

Typically, results in repeated measurement of a quantity θ_i are regarded as having been drawn from a normal distribution with mean θ_i. In the archetypal case, our knowledge of the variance of this distribution is confined to the spread of these results. If the number of results is small then the variance must be regarded as unknown. This is the situation envisaged in the measurement of the angle in Example 6.1, where we wrote (6.7) with σ^2 unknown. If Result 6.1 is to be applied in Example 6.1 then the parent distribution of the overall error $\hat{d} - d$ must be known, at least approximately. So somehow statement (6.7) must be replaced by one in which e_d is attributed a parent distribution that is fully specified. The following concepts are relevant.

Background

Let there be n_i raw measurement results $y_1, \ldots, y_{n_i}$ obtained when estimating a quantity θ_i. The observed sample mean is $\bar{y} = \sum_{j=1}^{n_i} y_j/n_i$ and the observed sample variance is

$$s_i^2 = \frac{\sum_{j=1}^{n_i}(y_j - \bar{y})^2}{n_i - 1}. \tag{6.8}$$

The estimate of θ_i is the sample mean, so

$$x_i = \bar{y}.$$

As is typical, let us suppose the errors in the individual y_i values have been independently drawn from a normal distribution with mean zero and unknown variance σ_i^2. The model for the generation of the error in x_i is therefore

$$x_i - \theta_i \leftarrow N(0, \sigma_i^2/n_i),$$

as in (6.7). It is well known that, although σ_i^2 remains unknown, a realized exact 95% confidence interval for θ_i under this model can be calculated using the t distribution with $n_i - 1$ degrees of freedom: the problem can be formulated in terms of a ratio of variates in which the 'nuisance parameter' σ_i^2 is absent. (See Example 3.1.) The realized interval is

$$[x_i - t_{n-1, 0.975} s_i/\sqrt{n_i}, \; x_i + t_{n-1, 0.975} s_i/\sqrt{n_i}],$$

6.4 Treating a normal error with unknown underlying variance

where $t_{\nu,0.975}$ denotes the 0.975 quantile of the t distribution with ν degrees of freedom. It is also known that if each θ_i is measured in this way then a realized approximate 95% confidence interval for $\sum a_i \theta_i$ is available using a t distribution with an effective number of degrees of freedom ν_{eff} given by (Willink, 2007a)

$$\nu_{\text{eff}} = \frac{\left(\sum a_i^2 s_i^2 / n_i\right)^2}{\sum \left(a_i^2 s_i^2 / n_i\right)^2 / (n_i - 1)}. \tag{6.9}$$

(See Example 3.2.) This interval is

$$\left[\sum a_i x_i - t_{\nu_{\text{eff}},0.975} \sqrt{\sum a_i^2 s_i^2 / n_i}, \ \sum a_i x_i + t_{\nu_{\text{eff}},0.975} \sqrt{\sum a_i^2 s_i^2 / n_i}\right]. \tag{6.10}$$

Using (6.9) and (6.10) seems to be the best approach to the analysis of measurement uncertainty when Type A errors of this form are dominant. Indeed, the method described in the *Guide* involves this approach (BIPM et al., 1995; JCGM, 2008a, Appendix G). However, the typical measurement problem will also involve various θ_i quantities that are not estimated in this way: in particular, one or more influential components of error will be of Type B. So the theory behind these results becomes inapplicable, and a theory involving the addition of Type A and Type B errors is required. In the statistical community there would be no agreed approach to the solution of this problem, one reason being that the attribution of known parent distributions to Type B errors is a controversial step. But, in measurement, this situation is ubiquitous and a practical solution must be found. There are at least three approaches that might be taken.

1. One possibility would be to treat Type A and Type B errors with proper regard for their differences. The goal would be to develop a confidence-procedure with success probability as little in excess of 95% as possible. (See Appendix D for a solution in the case of one Type A error and a uniform Type B error.) Even if this goal were feasible, the analysis might be overly complicated in the general case.
2. A second approach is to treat Type B errors in a way that appeals to the theory of treating Type A errors. In effect, this kind of approach is taken in the *Guide*, where a number of 'degrees of freedom' is associated with an evaluation of an uncertainty component by Type B methods. Subsequently an effective number of degrees of freedom is calculated using the 'Welch–Satterthwaite formula', which is a formula directly analogous to (6.9). There is much that is good in the procedure advocated in the *Guide* provided that it is interpreted as a frequentist procedure. The *Guide* is discussed in Chapter 14.

Final preparation

3. Conversely, the third approach is to treat the Type A errors in a way that fits Type B errors. This means attributing a known parent distribution to a Type A error. The parent distribution of the overall measurement error can then be calculated, and Results 6.1 and 6.2 become applicable.

In this book, we adopt the third approach. As stated, this approach permits the use of Results 6.1 and 6.2. It also facilitates simulation of the measurement process. Importantly, employing the distribution that we shall advocate for the Type A error does not seem to reduce the success rate below 95%.

Thus, we suggest attributing a known parent distribution to an error in a situation where the actual parent distribution is unknown. This is a step that would be unappealing to a statistical purist (like myself). The cost of taking this pragmatic step is statistical inefficiency in situations where the dominant errors are Type A errors based on small numbers of measurements, where using (6.9) and (6.10) with the recipe of the *Guide* would lead to narrower intervals.

Parent distribution for the error

We suggest that when θ_i is estimated by the mean x_i of n_i independent results, the parent distribution of the error $x_i - \theta_i$ be taken to be the distribution of $(s_i/\sqrt{n_i})T_{n_i-1}$, where s_i is given by (6.8) and T_ν is a variate with the t distribution with ν degrees of freedom. So the model is

$$x_i - \theta_i \leftarrow (s_i/\sqrt{n_i})t_{n-1},$$

in which ct_ν denotes the distribution of cT_ν. Before we describe the rationale for using this distribution, let us see how this step affects the model in Example 6.1.

Example 6.1 (continued) Measuring the distance d between the two points on the plane involves measuring the angle ϕ. This angle is estimated by averaging $n = 7$ readings each containing error drawn independently from a single normal distribution with mean zero. The sample standard deviation is found from (6.8) to be $s = 0.0108$ radian. The original model of the generation of the error in the estimate of angle, which is (6.7) and which involves the nuisance parameter σ^2, is now replaced by the model

$$e_d \leftarrow 0.00408\, t_6, \tag{6.11}$$

which involves no unknown parameters. This distribution is shown in Figure 6.2, along with two possibilities for the unknown actual parent distribution $N(\sigma^2/7)$.

6.4 Treating a normal error with unknown underlying variance

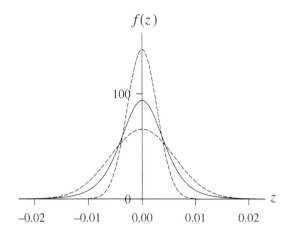

Figure 6.2 The parent distribution attributed to e_d, which is the known distribution $0.00408\, t_6$ (solid line) and two possibilities for the unknown parent distribution of e_d (dashed lines).

Our suggestion is that $(s_i/\sqrt{n_i})t_{n-1}$ be used as the parent distribution of the error $x_i - \theta_i$. There are several reasons for this choice of distribution. The first is that we obtain an exact confidence procedure when a single error of this type dominates all others, which is a condition approached in some simple measurement situations. To see this, observe that x_i is the outcome of a variate X_i that has the normal distribution with mean θ_i and variance σ_i^2/n_i, while the figure s_i^2 is the outcome of a variate S_i^2 distributed such that $(n_i - 1)S_i^2/\sigma_i^2$ has the chi-square distribution with $n_i - 1$ degrees of freedom. The variates X_i and S_i^2 are independent. Consequently, the variate $(X_i - \theta_i)/(S_i/\sqrt{n_i})$ has the t distribution with $n_i - 1$ degrees of freedom. (See Example 3.1.) This means that

$$\Pr\left(\left|\frac{X_i - \theta_i}{S_i/\sqrt{n_i}}\right| \le t_{n_i-1,0.975}\right) = 0.95$$

and, equivalently,

$$\Pr([X_i - t_{n_i-1,0.975}S_i/\sqrt{n_i},\ X_i + t_{n_i-1,0.975}S_i/\sqrt{n_i}] \ni \theta_i) = 0.95.$$

Therefore, the random interval

$$[X_i - t_{n_i-1,0.975}S_i/\sqrt{n_i},\ X_i + t_{n_i-1,0.975}S_i/\sqrt{n_i}]$$

is an exact 95% confidence interval for θ_i, and

$$[x_i - t_{n_i-1,0.975}s_i/\sqrt{n_i},\ x_i + t_{n_i-1,0.975}s_i/\sqrt{n_i}]$$

is a realized exact 95% confidence interval for θ_i. This is the same numerical interval that would be obtained using Result 6.1 under our suggestion that the parent distribution of the error $x_i - \theta_i$ be deemed to be $(s_i/\sqrt{n_i})t_{n_i-1}$.

The second reason for this choice is that the ensuing statistical error is small when n_i is large. The actual parent distribution of $x_i - \theta_i$ is the normal distribution with mean 0 and variance σ_i^2/n_i. The variate S_i^2 is a *consistent* estimator of σ_i^2, so in the limit of $n_i \to \infty$ the figure s_i^2 will be equal to σ_i^2. The t distribution approaches the standard normal distribution as the number of degrees of freedom increases, with equality when n_i is infinite. So when n_i is infinite, the parent distribution suggested for $x_i - \theta_i$ is the same as the actual parent distribution.

The third reason for this choice is that – other sources of approximation notwithstanding – it will lead to numerical intervals equivalent to those obtained in a typical 'objective Bayesian' analysis in which θ_i and σ_i are attributed the usual improper prior distributions. (See Chapter 13, in particular (13.1) and (13.2).) This will be of some appeal to those concerned with the standardization of analyses.

The fourth, and most important, reason is that this choice leads to a valid procedure. Other sources of error notwithstanding, the overall procedure that eventuates appears to have a success rate no less than 95%. To explain this idea we now assume a context where the total measurement error is the sum of independent components.

Let the measurement error $e = x - \theta$ be expressible as $\sum_{i=1}^{m} a_i e_i$, where each a_i is a known coefficient and each e_i is the outcome of an independent error variate E_i. So e is the outcome of a variate $E = \sum_{i=1}^{m} a_i E_i$. Let the E_i variates be indexed such that the first k are of Type A and the remaining $m - k$ are of Type B. Suppose that the distribution of a Type A error is $N(0, \sigma_i^2/n_i)$ with σ_i unknown, and suppose that the distribution of each Type B error is symmetric. If we treat the Type A errors in the manner proposed then $|E|$ is regarded as having the distribution of

$$Q \equiv \left| \sum_{i=1}^{k} a_i (s_i/\sqrt{n_i}) T_{n_i-1} + \sum_{i=k+1}^{m} a_i E_i \right|.$$

The measurement is successful when the modulus of the error incurred does not exceed the half-length of the 95% interval generated for θ. This half-length is the 0.95 quantile of the distribution of Q, which is a figure we denote by $q_{0.95}(s_1, \ldots, s_k)$. The figure s_i is the outcome of a variate S_i that has a density function $f_{S_i}(z_i)$ defined by σ_i and n_i. So the probability of success is

$$\Pr(\text{success}) = \int_0^\infty \cdots \int_0^\infty \Pr\{|E| \leq q_{0.95}(z_1, \ldots, z_k)\} f_{S_1}(z_1) \cdots f_{S_k}(z_k) \, dz_k \cdots dz_1,$$

the variates $E, S_1, \ldots, S_k$ being independent. Our conjecture is that this probability is at least 0.95 in every situation, each situation being defined by the values of $m, k, \{\sigma_1, \ldots, \sigma_k\}, \{n_1, \ldots, n_k\}$ and by the distributions of $E_{k+1}, \ldots, E_m$.

6.4 Treating a normal error with unknown underlying variance

We have already seen that the probability of success is exactly 0.95 when $k = m = 1$ or when each n_i is infinite. The same is true when all the error components are of Type B, which is when $k = 0$. Also, because S_i^2 is an unbiased estimator of σ_i^2 and the variance of a t distribution is at least unity, it can be shown that in the limit of $k \to \infty$ the variance of the parent distribution calculated for the measurement error is not less than the variance of the actual distribution, in which case the probability of success will be at least 0.95.

Thus, there are a number of special cases in which the performance of the method is consistent with our conjecture that the procedure errs on the side of conservatism, so that the procedure is valid. However, proving this result for the general case seems difficult, and so we rely on the results of Monte Carlo analysis. Appendix E summarizes a Monte Carlo analysis that strongly supports the conjecture. In fact, the tendency for the procedure to be conservative when Type A errors are influential seems to be its main drawback. When the total error is dominated by Type A errors, it will lead to wider intervals than would be obtained using the approach based on an effective number of degrees of freedom.

Summary

At last, our preparation for the description of the evaluation of measurement uncertainty is complete. An important part of this preparation is the step just taken of attributing a known distribution to a Type A error variate after observation of the data, which is a step that a theoretical classical statistician would be loathe to take. This is a pragmatic step taken because of the inevitability of systematic errors in measurement problems. From a theoretical point of view, this step seems preferable to the step of treating non-normal Type B errors as if they were normal Type A errors, which is the idea behind the approach taken in the *Guide* (1995, 2008a). It enables the use of Results 6.1 and 6.2, and it also facilitates approximate simulation of the measurement by the Monte Carlo method. These ideas will be seen in action in the chapters to follow.

7

Evaluation using the linear approximation

We are now in a position to describe different methods for the evaluation of measurement uncertainty. This chapter discusses methods that involve the well-known step of using a linear function to approximate the measurand equation in the vicinity of the observation point $(x_1, \ldots, x_m)$. Section 7.1 describes this linear approximation and the associated laws of propagation of error and uncertainty. Section 7.2 gives a familiar analytical method for the evaluation of measurement uncertainty when the covariances of the E_i variates are known and where the total error E is approximately normal. Section 7.3 describes a more accurate method applicable when the total error can be expressed as the sum of independent components with known moments. Section 7.4 then shows how the Monte Carlo principle might be used to obtain a method with yet greater accuracy.

7.1 Linear approximation to the measurand equation

Recall that the target value is expressible in terms of various other unknown quantities by the measurand equation $\theta = \mathcal{F}(\theta_1, \ldots, \theta_m)$. Consider a fixed point $\mathbf{b} \equiv (b_1, \ldots, b_m)$ in the vicinity of the unknown point $(\theta_1, \ldots, \theta_m)$. In practice, $\mathcal{F}$ will possess a Taylor's series expansion about $\mathbf{b}$, in which case we can write

$$\theta = \mathcal{F}(\mathbf{b}) + \sum_{i=1}^{m}(\theta_i - b_i)\frac{\partial \mathcal{F}(\mathbf{b})}{\partial b_i} + \cdots \qquad (7.1)$$

with

$$\frac{\partial \mathcal{F}(\mathbf{b})}{\partial b_i} \equiv \left.\frac{\partial \mathcal{F}(b_1, \ldots, b_{i-1}, z, b_{i+1}, \ldots, b_m)}{\partial z}\right|_{z=b_i}.$$

Truncating after the last term shown in (7.1) gives a linear approximation to the measurand function.

7.1 Linear approximation to the measurand equation

In the textbook situation, each θ_i is a parameter and each x_i is the outcome of a variate X_i. The variate for the measurement result is then $X = \mathcal{F}(X_1, \ldots, X_m)$, which can similarly be written as

$$X = \mathcal{F}(\mathbf{b}) + \sum_{i=1}^{m}(X_i - b_i)\frac{\partial \mathcal{F}(\mathbf{b})}{\partial b_i} + \cdots.$$

It follows that

$$X = \theta + \sum_{i=1}^{m}(X_i - \theta_i)\frac{\partial \mathcal{F}(\mathbf{b})}{\partial b_i} + \cdots.$$

This equation holds for all points $\mathbf{b}$ and so it holds for the point $\mathbf{x} \equiv (x_1, \ldots, x_m)$ even though we do not know this point until the variates $X_1, \ldots, X_m$ have been realized. Therefore, an approximation that is correct to terms of first order is

$$X \approx \theta + \sum_{i=1}^{m} E_i \frac{\partial \mathcal{F}(\mathbf{x})}{\partial x_i}, \tag{7.2}$$

where $E_i = X_i - \theta_i$. The analogous postmeasurement equation is

$$x \approx \theta + \sum_{i=1}^{m} e_i \frac{\partial \mathcal{F}(\mathbf{x})}{\partial x_i}, \tag{7.3}$$

which can be written as

$$\theta \approx x - \sum_{i=1}^{m} e_i \frac{\partial \mathcal{F}(\mathbf{x})}{\partial x_i}. \tag{7.4}$$

Expressions (7.3) and (7.4) might be called the 'linear approximation to the estimate equation' and the 'linear approximation to the measurand equation' respectively. We will use the general term *linear approximation* to refer to this approach.

Equation (7.2) describes an approximation to the estimator X of θ. Result 6.1 then implies that a realized approximate 95% confidence interval for θ is

$$[x - e'_{0.975},\ x - e'_{0.025}], \tag{7.5}$$

where $e'_{0.025}$ and $e'_{0.975}$ are the 0.025 and 0.975 quantiles of the parent distribution of $e' \equiv \sum_{i=1}^{m} e_i\, \partial \mathcal{F}(\mathbf{x})/\partial x_i$, which is the distribution of

$$E' \equiv \sum_{i=1}^{m} E_i \frac{\partial \mathcal{F}(\mathbf{x})}{\partial x_i}. \tag{7.6}$$

These quantiles can be found to a level of accuracy that depends on our knowledge of the parent distributions of each component of error, e_i.

Example 6.1 (continued) The distance d between two points in the plane is measured. (See Figure 6.1.) The estimate equation is

$$\hat{d} = \sqrt{\hat{r}_1^2 + \hat{r}_2^2 - 2\hat{r}_1\hat{r}_2 \cos\hat{\phi}}. \tag{7.7}$$

Therefore,

$$\frac{\partial \hat{d}}{\partial \hat{r}_1} = \frac{\hat{r}_1 - \hat{r}_2 \cos\hat{\phi}}{\hat{d}},$$

$$\frac{\partial \hat{d}}{\partial \hat{r}_2} = \frac{\hat{r}_2 - \hat{r}_1 \cos\hat{\phi}}{\hat{d}},$$

$$\frac{\partial \hat{d}}{\partial \hat{\phi}} = \frac{\hat{r}_1 \hat{r}_2 \sin\hat{\phi}}{\hat{d}}$$

and the measurement error under the linear approximation is

$$e' = \frac{1}{\hat{d}}\left\{(\hat{r}_1 - \hat{r}_2 \cos\hat{\phi})(\hat{r}_1 - r_1) + (\hat{r}_2 - \hat{r}_1 \cos\hat{\phi})(\hat{r}_2 - r_2) - \hat{r}_1\hat{r}_2(\hat{\phi} - \phi)\right\}. \tag{7.8}$$

Suppose the estimates of the three unknowns are $\hat{r}_1 = 54.8$ m, $\hat{r}_2 = 78.2$ m and $\hat{\phi} = 0.484$ rad. The estimate of the distance between the two points is 39.14 m, and the partial derivatives are

$$\frac{\partial \hat{d}}{\partial \hat{r}_1} = -0.368,$$

$$\frac{\partial \hat{d}}{\partial \hat{r}_2} = 0.759,$$

$$\frac{\partial \hat{d}}{\partial \hat{\phi}} = 50.95 \text{ m rad}^{-1}.$$

The parent distributions of the errors $\hat{r}_1 - r_1$, $\hat{r}_2 - r_2$ and $\hat{\phi} - \phi$ were described in Sections 6.3 and 6.4. We will see in Section 7.2 that the appropriate quantiles of the corresponding parent distribution of e' are $e'_{0.025} \approx -0.55$ m and $e'_{0.975} \approx 0.55$ m. Therefore, from (7.5), a realized approximate 95% confidence interval for d is

$$[38.59, 39.69] \text{ m}. \tag{7.9}$$

There is no relevant information unused in the calculation of this interval, so this interval can be taken as a 95% uncertainty interval for θ.

Calculation of partial derivatives

Analytical calculation of the partial derivatives might be difficult in some cases. However, a straightforward numerical method exists. Let δ be equal to a small fraction of $\sigma_p(e_i)$, say one tenth of the recorded unit of resolution for x_i. Then

7.1 Linear approximation to the measurand equation

$$\frac{\partial \mathcal{F}(\mathbf{x})}{\partial x_i} \approx \frac{\mathcal{F}(x_1, \ldots, x_{i-1}, x_i + \delta, x_{i+1}, \ldots, x_m) - x}{\delta} \quad (7.10)$$

with high accuracy. In many situations, the calculation of derivatives by this method will be easier than calculation analytically.

Example 6.1 (continued) The distance d between the two points in the plane is measured. The estimate equation is (7.7). The relevant estimates are $\hat{r}_1 = 54.8$ m, $\hat{r}_2 = 78.2$ m, $\hat{\phi} = 0.484$ rad and $\hat{d} = 39.14$ m. Setting $\delta = 0.01$ m for the radial distances and $\delta = 0.0001$ rad for the angle and applying (7.10) gives the results

$$\frac{\partial \hat{d}}{\partial \hat{r}_1} \approx -0.368,$$

$$\frac{\partial \hat{d}}{\partial \hat{r}_2} \approx 0.759,$$

$$\frac{\partial \hat{d}}{\partial \hat{\phi}} \approx 50.95 \text{ m rad}^{-1},$$

which agree with the results obtained analytically.

The law of propagation of error and uncertainty

The linear approximation gives rise to 'the law of propagation of error'. The variate for the measurement error is $E = X - \theta$. From the relationship $E \approx E'$ and from (7.6) we find that

$$\sigma^2(E) \approx \sum_{i=1}^{m} \sum_{j=1}^{m} \frac{\partial \mathcal{F}(\mathbf{x})}{\partial x_i} \frac{\partial \mathcal{F}(\mathbf{x})}{\partial x_j} \text{cov}(E_i, E_j), \quad (7.11)$$

with the quantity on the right-hand side of this equation being the variance of E'. An equivalent equation is

$$\sigma^2(E) \approx \sum_{i=1}^{m} \sum_{j=1}^{m} \frac{\partial \mathcal{F}(\mathbf{x})}{\partial x_i} \frac{\partial \mathcal{F}(\mathbf{x})}{\partial x_j} \rho(E_i, E_j) \sigma(E_i) \sigma(E_j), \quad (7.12)$$

with $\rho(E_i, E_j)$ being the correlation coefficient between E_i and E_j. When the error variates are uncorrelated we obtain

$$\sigma^2(E) \approx \sum_{i=1}^{m} \left(\frac{\partial \mathcal{F}(\mathbf{x})}{\partial x_i}\right)^2 \sigma^2(E_i).$$

These equations represent the law of propagation of error, but a more accurate title would be 'the law of propagation of error variance'.

Sometimes the only dependence between pairs of errors will be perfect, in which case each correlation coefficient in (7.12) will be 0, 1 or −1. In most other cases, the correlation coefficients will be derived from the covariances. So (7.11) is a more fundamental equation than (7.12), and it will often be an easier equation with which to work.

Errors exist whether we know their statistical properties or not. So the law of propagation of error is meaningful whether the covariances in (7.11) are known or not. In contrast, measurement uncertainty is something that is known, not estimated. So any law about propagating measurement uncertainty must only involve quantities that are known and must not be an approximation. Therefore, it must be written in terms of known estimates of the covariances. The corresponding estimate of $\sigma^2(E)$ is given by

$$\hat{\sigma}^2(E) = \sum_{i=1}^{m} \sum_{j=1}^{m} \frac{\partial \mathcal{F}(\mathbf{x})}{\partial x_i} \frac{\partial \mathcal{F}(\mathbf{x})}{\partial x_j} \widehat{\mathrm{cov}}(E_i, E_j).$$

Similarly, the equation analogous to (7.12) must involve estimates of the standard deviations and correlation coefficients. It will be

$$\hat{\sigma}^2(E) = \sum_{i=1}^{m} \sum_{j=1}^{m} \frac{\partial \mathcal{F}(\mathbf{x})}{\partial x_i} \frac{\partial \mathcal{F}(\mathbf{x})}{\partial x_j} \hat{\rho}(E_i, E_j) \hat{\sigma}(E_i) \hat{\sigma}(E_j).$$

If each E_i has mean zero then E' has mean zero. In this situation the measurement uncertainty will be some multiple of $\hat{\sigma}(E)$, in which case these equations both seem well described as the 'law of propagation of (measurement) uncertainty'.

The law of propagation of error and the law of propagation of uncertainty are not emphasized as much as might be expected in this book. They relate chiefly to the calculation of variances, whereas our focus is on the calculation of uncertainty intervals. These laws are insufficient for the calculation of an uncertainty interval unless more is known or assumed about the distribution of E'.

Incorporating fixed estimates

So far, our analysis in this chapter has been for the standard situation where each θ_i is a parameter of the measurement and each x_i is the outcome of a variate X_i. However, in many measurements one or more of the x_i values will be fixed, as in Example 1.1 where $\hat{\delta} = 0$ and the corresponding error was a matching error. If x_i is fixed then the concept of 'repetition' that gives practical meaning to the idea of probability is the concept of carrying out many measurements of this type with different values of θ_i spread around x_i. Consequently, we must see θ_i as being the realization of a variate Θ_i that would take different values in these different measurements.

In this situation, (7.2) and (7.6) hold provided that E_i is defined to be $x_i - \Theta_i$ when x_i is fixed. Similarly (7.3) and (7.4) hold. The interval estimator of θ, which is the random interval $[X - e'_{0.975}, X - e'_{0.025}]$, becomes an average confidence interval, as described in Section 5.1. The quoted figure of probability, 0.95, then refers to the success rate over the hypothetical set of measurements in which θ_i is distributed around x_i. So, when using the linear approximation to obtain a 95% interval of measurement uncertainty, we do not need to make special allowances for the fact that one or more of the x_i estimates might be fixed. The only requirement is that the interval (7.5) be understood to be the outcome of an average confidence interval, not the outcome of an ordinary confidence interval.

Also, one or more of the θ_i values might not be a parameter of the measurement but might be created during the measurement process. For example, one of the θ_i quantities might be an uncontrolled environmental quantity like temperature. In such a case, (7.2)–(7.6) hold, but the interval (7.5) should then be understood to be the outcome of a conditional confidence interval, as described in Section 5.2.

7.2 Evaluation assuming approximate normality

The law of propagation of error is not always sufficient for the calculation of a confidence interval. It is, however, sufficient when the total error is normal and the error covariances are known, or when these things hold approximately. Using postmeasurement notation, we can write (7.11) as $\sigma_p^2(e) \approx \sigma_p^2(e')$ with

$$\sigma_p^2(e') = \sum_{i=1}^{m} \sum_{j=1}^{m} \frac{\partial \mathcal{F}(\mathbf{x})}{\partial x_i} \frac{\partial \mathcal{F}(\mathbf{x})}{\partial x_j} \operatorname{cov}_p(e_i, e_j). \qquad (7.13)$$

If the parent distribution of e' is normal then a realized approximate 95% confidence interval for θ is

$$[x - 1.96\sigma_p(e'), x + 1.96\sigma_p(e')].$$

The interval is approximate because we have made the linear approximation. In practice, it will be also be approximate because the parent distribution of e' will not be exactly normal, perhaps because we are relying on the action of a central limit theorem. Nevertheless, the result can be called a 95% uncertainty interval for θ unless there is external information to call into question its reliability.

Example 6.1 (continued) The distance d between the two points in the plane is measured. The data are $\hat{r}_1 = 54.8$ m, $\hat{r}_2 = 78.2$ m and $\hat{\phi} = 0.484$ rad, and the measurement estimate is $\hat{d} = 39.14$ m. The measurement error under the linear approximation is given by (7.8) and the error model is

104   Evaluation using the linear approximation

Table 7.1. *Estimate of the parent covariance matrix of the errors*

Error	$\hat{r}_1 - r_1$ (m)	$\hat{r}_2 - r_2$ (m)	$\hat{\phi} - \phi$ (rad)
$\hat{r}_1 - r_1$	0.022	0.017	0
$\hat{r}_2 - r_2$	0.017	0.034	0
$\hat{\phi} - \phi$	0	0	2.50×10^{-5}

$$\hat{r}_1 - r_1 = r_1 e_a + e_b,$$
$$\hat{r}_2 - r_2 = r_2 e_a + e_c,$$
$$\hat{\phi} - \phi = e_d,$$
$$e_a \leftarrow D\left(0, 0.002^2\right),$$
$$e_b \leftarrow D(0, 0.1^2),$$
$$e_c \leftarrow D(0, 0.1^2),$$
$$e_d \leftarrow 0.00408 \, t_6 \qquad \text{(which is (6.11))},$$

e_a, e_b, e_c, e_d have been drawn independently.

The variance of the distribution t_ν is $\nu/(\nu - 2)$. So the parent variances of the errors in the three estimates are

$$\text{var}_p(\hat{r}_1 - r_1) = 0.002^2 \, r_1^2 + 0.1^2, \tag{7.14}$$
$$\text{var}_p(\hat{r}_2 - r_2) = 0.002^2 \, r_2^2 + 0.1^2, \tag{7.15}$$
$$\text{var}_p(\hat{\phi} - \phi) = 0.00408^2 \times 6/(6 - 2) = 2.50 \times 10^{-5}$$

and the parent covariance of the errors in $\hat{r}_1$ and $\hat{r}_2$ is

$$\text{cov}_p\left(\hat{r}_1 - r_1, \hat{r}_2 - r_2\right) = 4 r_1 r_2 \times 10^{-6}. \tag{7.16}$$

The unknowns r_1 and r_2 on the right-hand sides of (7.14)–(7.16) are approximated by their estimates. We obtain the estimate of the parent covariance matrix given in Table 7.1.

Applying (7.13) shows that the parent variance of e' is given by

$$\sigma_p^2(e') \approx 0.0781 \, \text{m}^2, \tag{7.17}$$

so the parent standard deviation of e' is $\sigma_p(e') \approx 0.279$ m. Although the parent distributions of e_a, e_b and e_c are unspecified, we suppose that the parent distribution of e' is close to normal through the action of the central limit theorem for non-identical variables. Then $-e'_{0.025} \approx e'_{0.975} \approx 1.960 \, \sigma_p(e') \approx -0.548$ m, and we obtain the interval [38.59, 39.69] m, which is (7.9).

The procedure of this section will be applicable in many situations. Indeed, some measurement scientists might argue that it is the only procedure needed. We shall refer to it as the *normal-approximation method*.

7.3 Evaluation using higher moments

Often the total 'linearized' error variate E' of (7.6) will be decomposable into independent additive variates, these being either the $E_i \, \partial \mathcal{F}(\mathbf{x})/\partial x_i$ contributions or smaller underlying components. In this situation, an alternative method of evaluation based on combining the moments of the contributing variates is available. It involves the useful idea that when independent variates are added their cumulants are added. The relevant theory is as follows.

Theory 7.1 The set of cumulants of a variate, Y, is a transform of its set of moments. The first cumulant of Y is its first moment about the origin, which is the mean $\mathcal{E}(Y)$. The second cumulant of Y is its variance, which is the second central moment $\sigma^2(Y) \equiv \mathcal{E}[(Y - \mathcal{E}(Y))^2]$. The third cumulant of Y is the third central moment $\mathcal{E}[(Y - \mathcal{E}(Y))^3]$, which is zero if the distribution is symmetric. The fourth cumulant of Y is

$$\kappa_4(Y) = \mu_4(Y) - 3\sigma^4(Y), \qquad (7.18)$$

where $\mu_4(Y)$ is the fourth central moment, i.e. $\mu_4(Y) = \mathcal{E}[(Y - \mathcal{E}(Y))^4]$.

The dimensionless ratio

$$\gamma(Y) \equiv \frac{\kappa_4(Y)}{\sigma^4(Y)} \qquad (7.19)$$

is called the *coefficient of excess* of Y. It is a function of the shape of the distribution only, and it measures the excess kurtosis over that of a normal distribution. The coefficient of excess of a normal variable is zero (because a normal variable has rth cumulant equal to zero for $r \geq 3$). The coefficient of excess of a symmetric variable is positive if the distribution is symmetric with heavier tails, such as a t distribution, and is negative if the distribution is symmetric with blunter tails, such as a uniform distribution. The minimum possible coefficient of excess is -2, which is attained when the distribution has the 'maximal variance' form of Figure 4.5.

Table 7.2 shows the variances, fourth cumulants and coefficients of excess of several symmetric distributions. These are the Laplace distribution with standard deviation c, denoted $L(c)$; the distribution of c times a variate with the t distribution with ν degrees of freedom, ct_ν; the normal distribution with zero mean and variance c^2, $N(0, c^2)$; the isosceles triangular distribution on the interval from $-c$ to c, denoted $I(-c, c)$; the uniform distribution on the same interval, $U(-c, c)$; and

Table 7.2. *Variance σ^2, fourth cumulant κ_4 and coefficient of excess γ of symmetric distributions*

Distribution		σ^2	κ_4	γ
Laplace	L(c)	$2c^2$	$12c^4$	3
Scaled t	ct_ν	$\dfrac{c^2\nu}{\nu-2}$	$\dfrac{6c^4\nu^2}{(\nu-2)^2(\nu-4)}$	$\dfrac{6}{\nu-4}$
Normal	N(0, c^2)	c^2	0	0
Triangular	I($-c, c$)	$c^2/6$	$-c^4/60$	-0.6
Uniform	U($-c, c$)	$c^2/3$	$-2c^4/15$	-1.2
Arc-sine	A($-c, c$)	$c^2/2$	$-3c^4/8$	-1.5

the arc-sine distribution on the same interval, denoted A($-c, c$). In each case c is simply a scale factor.

The notation μ_r and κ_r for the rth central moment and rth cumulant is standard. Our notation γ for the coefficient of excess is non-standard, a more familiar notation in statistics being γ_2. We omit the subscript for simplicity.

The technique of uncertainty analysis that we now propose requires that the first four moments of each component error variate be known. This means that we know the first four cumulants of each component. The description given here relates to the usual situation where each component is symmetric and has mean zero, so its first and third cumulants are zero. Each component is then represented by its variance and fourth cumulant, or by its variance and coefficient of excess. The parent distribution for the overall measurement error will also be symmetric and will have mean zero. Notwithstanding the effect of using the linear approximation $E \approx E'$, the variance and coefficient of excess of the overall error variate E can be obtained by applying the following theory to the terms in the equation

$$E' = \sum_{i=1}^{m} E_i \frac{\partial \mathcal{F}(\mathbf{x})}{\partial x_i},$$

which is (7.6).

Theory 7.1 (continued) The rth cumulant of the variate aY is a^r times the rth cumulant of Y. And if $Y_1, \ldots, Y_n$ are independent variates then the rth cumulant of $\sum_{j=1}^{n} Y_j$ is the sum of the rth cumulants of the Y_j variates (Johnson et al., 1993, p. 45). It follows that if $Y_1, \ldots, Y_n$ are independent then

$$\sigma^2 \left(\sum_{j=1}^{n} a_j Y_j \right) = \sum_{j=1}^{n} a_j^2 \sigma^2(Y_j)$$

and

$$\kappa_4\left(\sum_{j=1}^n a_j Y_j\right) = \sum_{j=1}^n a_j^4 \kappa_4(Y_j),$$

the first of these two results being familiar. It is then easily shown that

$$\gamma\left(\sum_{j=1}^n a_j Y_j\right) = \frac{\sum_{j=1}^n a_j^4 \gamma(Y_j)\sigma^4(Y_j)}{\left\{\sum_{j=1}^n a_j^2 \sigma^2(Y_j)\right\}^2}.$$

The results given in Theory 7.1 are directly applicable to the task of approximating the distribution of E'. Let E_i be symmetric with mean 0, variance $\sigma^2(E_i)$ and coefficient of excess $\gamma(E_i)$. Then, if each E_i is independent, E' has mean 0, variance

$$\sigma^2(E') = \sum_{i=1}^m \left(\frac{\partial \mathcal{F}(\mathbf{x})}{\partial x_i}\right)^2 \sigma^2(E_i)$$

and coefficient of excess

$$\gamma(E') = \frac{\sum_{i=1}^m (\partial \mathcal{F}(\mathbf{x})/\partial x_i)^4 \gamma(E_i)\sigma^4(E_i)}{\sigma^4(E')}.$$

Evidently, E' will also be symmetric.

The figures $\sigma^2(E')$ and $\gamma(E')$ tell us about the scale and shape of E'. The distribution of E' can be approximated by another symmetric distribution with the same mean (zero), the same variance and the same coefficient of excess, which is a distribution with the same first four moments. We define our approximation to be the corresponding distribution from the Pearson system of distributions.

The Pearson system of distributions

The important system of distributions derived by Karl Pearson in the 1890s describes variates for which the probability density function $f(z)$ satisfies $f'(z)/f(z) = (z - a_1)/(a_2 + a_3 z + a_4 z^2)$ for constants $a_1, \ldots, a_4$ (Stuart and Ord, 1987; Johnson et al., 1994). It includes all 'linear transformations' of most of the well-known continuous distributions, including the normal, t, chi-square, uniform, gamma, beta, arc-sine and F distributions. Figure 7.1 shows various *standardized* symmetric Pearson distributions, i.e. those with mean 0 and variance 1. In particular, the value of $\gamma = 6$ gives a scaled t distribution with five degrees of freedom, $\gamma = 0$ gives the standard normal distribution, $\gamma = -1.2$ gives a uniform distribution and $\gamma = -1.5$ gives an arc-sine distribution.

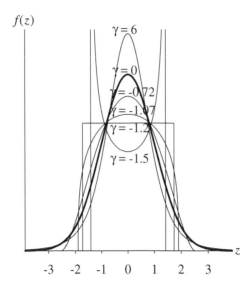

Figure 7.1 Standardized symmetric Pearson distributions with coefficients of excess indicated. (Normal distribution – thick line.)

There is one and only one Pearson distribution for each possible set of the first four moments, and this distribution changes with the moments in a continuous way. So there is a unique symmetric Pearson distribution with mean zero, variance $\sigma^2(E')$ and coefficient of excess $\gamma(E')$. If $\gamma = 0$ then this distribution is normal, if $\gamma > 0$ it is the distribution of a multiple of a variable with a t distribution and if $\gamma < 0$ it is the distribution of a multiple of a variable with a symmetric beta distribution shifted to have mean zero.

The 0.975 and 0.995 quantiles of the standardized symmetric Pearson distribution with coefficient of excess γ between -1.2 and 6 are shown in Figure 7.2. These quantiles are very closely approximated by the figures

$$k_{0.975} = \frac{1.96 + 1.845\,\gamma + 0.47\,\gamma^2}{1 + 0.906\,\gamma + 0.239\,\gamma^2} \tag{7.20}$$

and

$$k_{0.995} = \frac{2.5758 + 2.6736\,\gamma + 0.7685\,\gamma^2}{1 + 0.8864\,\gamma + 0.2362\,\gamma^2}, \tag{7.21}$$

the values for $\gamma = 0$ being recognizable as the 0.975 and 0.995 quantiles of the standard normal distribution. These approximations have relative error within $\pm 0.025\%$ for $\gamma \in \{-1.20, -1.19, \ldots, 6.00\}$ but they are also useful over a wider range of values of γ. The figure $k_{0.975}$ has relative error within $\pm 1\%$ for

7.3 Evaluation using higher moments

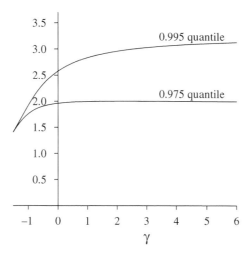

Figure 7.2 The 0.975 and 0.995 quantiles of the standardized symmetric Pearson distribution with coefficient of excess γ.

$\gamma \in \{-1.50, -1.49, \ldots, 100.00\}$, while the figure $k_{0.995}$ has relative error within $\pm 1\%$ for $\gamma \in \{-1.38, -1.37, \ldots, 100.00\}$.

Calculation of the uncertainty interval

Our analysis leads to the following result.

Result 7.1 Suppose that the target value is $\theta = \mathcal{F}(\theta_1, \ldots, \theta_m)$ and that x_i is an estimate of θ_i. The error $e_i = x_i - \theta_i$ is regarded as the outcome of a variate E_i. Let the distribution of E_i be symmetric with mean 0, variance $\sigma^2(E_i)$ and coefficient of excess $\gamma(E_i)$, and let each E_i be independent. Set $x = \mathcal{F}(x_1, \ldots, x_m)$,

$$\sigma^2 = \sum_{i=1}^{m} \left(\frac{\partial \mathcal{F}(\mathbf{x})}{\partial x_i} \right)^2 \sigma^2(E_i)$$

and

$$\gamma = \sigma^{-4} \sum_{i=1}^{m} \left(\frac{\partial \mathcal{F}(\mathbf{x})}{\partial x_i} \right)^4 \gamma(E_i) \sigma^4(E_i).$$

If $-1.5 \leq \gamma \leq 100$ then a realized approximate 95% confidence interval for θ is $[x - k_{0.975}\sigma, x + k_{0.975}\sigma]$, with $k_{0.975}$ being given in (7.20).

The use of this result is illustrated in the following example.

Example 7.1 The length L of an object is measured using three separate devices. The estimates obtained are $\hat{L}_1 = 81.21$, $\hat{L}_2 = 81.98$ and $\hat{L}_3 = 81.35$ in some units. In each case the dominant component of error is a Type B error modelled as having been drawn from a uniform distribution with mean zero; the half-widths of this distribution being $a_1 = 0.8$, $a_2 = 1.2$ and $a_3 = 0.4$ in the three measurements respectively. All other sources of error can be neglected.

The final estimate of L is a weighted mean of the three individual estimates $\hat{L}_1$, $\hat{L}_2$ and $\hat{L}_3$, the weights being chosen to minimize the parent variance of the overall error. When such components of error are independent, their optimal weights are inversely proportional to their parent variances (e.g. Hodges and Lehmann (1964), Sections 10.1, 10.4). In this case, these variances are proportional to the squares of the a_j values. So the final estimate of the length is

$$\hat{L} = \frac{\sum_{j=1}^{3} \hat{L}_j/a_j^2}{\sum_{j=1}^{3} 1/a_j^2} = 81.38. \qquad (7.22)$$

The corresponding error $e = \hat{L} - L$ is

$$e = \sum_{j=1}^{3} \frac{\partial \hat{L}}{\partial \hat{L}_j} e_j,$$

with $e_j = \hat{L}_j - L_j$ and

$$\frac{\partial \hat{L}}{\partial \hat{L}_j} = \frac{1/a_j^2}{\sum_{j=1}^{3} 1/a_j^2}.$$

Because $e_j \leftarrow U(-a_j, a_j)$ we see from Table 7.2 that

$$\sigma_p^2(e_j) = a_j^2/3, \qquad j = 1, 2, 3,$$
$$\gamma_p(e_j) = -1.2, \qquad j = 1, 2, 3.$$

Applying Result 7.1 then gives

$$\sigma_p^2(e) = 0.039,$$
$$\gamma_p(e) = -0.70,$$
$$k_{0.975} = 1.86$$

and shows that a realized approximate 95% confidence interval for θ is [81.01, 81.74]. This interval is shorter than the interval [80.99, 81.76] that would be obtained using the normal-approximation method because the parent distribution of the total error, being the convolution of three uniform distributions, has a negative coefficient of excess.

Figure 7.3 shows the parent probability density function of e as evaluated by Monte Carlo analysis using 5×10^6 trials, as approximated by the symmetric Pearson

7.3 Evaluation using higher moments

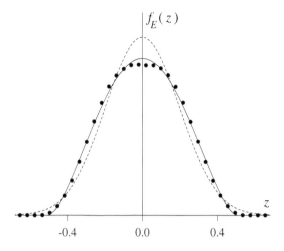

Figure 7.3 Dots: parent probability density function of e, i.e probability density function of E, as derived by Monte Carlo evaluation. Solid line: approximating Pearson distribution. Dashed line: approximating normal distribution.

distribution, and as approximated by the normal distribution with the same variance. The Pearson distribution is seen to provide a good approximation.

This example exhibits several peculiarities. First, every component error distribution happens to have the same form. Second, the problem is linear, so $e = e'$. Third, the measurand equation $\theta = \mathcal{F}(\theta_1, \ldots, \theta_m)$ has not been made explicit. From (7.22), we can take the measurand equation to be the identity

$$L = \frac{\sum_{j=1}^{3} L_j / a_j^2}{\sum_{j=1}^{3} 1/a_j^2}$$

with $L_1 = L_2 = L_3 = L$. None of these peculiarities affects the applicability of the method.

Sources of inaccuracy – numerical and statistical

Any non-zero mean in a Type B error variate is a bias that will be taken into account when forming the estimate, so the requirement that the distribution of each Type B error has mean zero is not restrictive. The numerical accuracy of the figure $k_{0.975}\sigma$ in Result 7.1 as an approximation to the 0.975 quantile of E' depends on how closely the selected Pearson distribution matches the distribution of E' in the tails. The Pearson distribution can be expected to give a better approximation than the normal distribution because the Pearson system contains the normal distribution as a special case.

112   *Evaluation using the linear approximation*

In statistics, the term *accuracy* refers to the similarity between the stated probability and the actual rate. So the accuracy of a confidence-interval procedure refers to its success rate, not to the numerical limits arising in any particular application of that procedure. And the practical success rate of a procedure relates to application in a series of different problems, not hypothetical repeated application in the same problem. During such a series of problems, we can expect the numerical error in the figure $k_{0.975}\sigma$ to take different signs, so the effect of such error on the success rate will be mitigated. To some extent, the same will be true for the error caused by the use of the linear approximation.

Dependent E_i error variates

We now show that the method is applicable with dependent E_i variates if these variates can be written in terms of independent components with known moments. First, a comment about notation is required.

Throughout this book we use the italicized subscript i as the numerical index for an error e_i and for its corresponding quantities θ_i and x_i. However, in Example 6.1 we used an upright alphabetic subscript for an underlying *independent* component of error, e.g. we used e_a to denote the relative error when measuring from the origin. We will continue to use upright alphabetic subscripts when we want to emphasize that the errors are independent, and we will indicate summation over all alphabetic subscripts by summing over the subscript 'w'.

The following result is a corollary of Result 7.1 written in postmeasurement notation. It is applicable when each e_i can be written as a sum of components $e_\mathrm{a}, e_\mathrm{b}, e_\mathrm{c}, \ldots$ that were drawn independently.

Result 7.2 Let x be an estimate of the target value θ. Suppose that the error $x - \theta$ can be approximated as the sum of components drawn from independent zero-mean symmetric distributions according to the equation

$$x - \theta \approx \sum_\mathrm{w} a_\mathrm{w} e_\mathrm{w},$$

where each a_w is known. Let the parent distribution of e_w have variance $\sigma_\mathrm{p}^2(e_\mathrm{w})$ and coefficient of excess $\gamma_\mathrm{p}(e_\mathrm{w})$. Define

$$\sigma^2 = \sum_\mathrm{w} a_\mathrm{w}^2 \sigma_\mathrm{p}^2(e_\mathrm{w})$$

and

$$\gamma = \frac{\sum_\mathrm{w} a_\mathrm{w}^4 \gamma_\mathrm{p}(e_\mathrm{w}) \sigma_\mathrm{p}^4(e_\mathrm{w})}{\sigma^4}.$$

7.3 Evaluation using higher moments

If $-1.5 \leq \gamma \leq 100$ then a realized approximate 95% confidence interval for θ is $[x - k_{0.975}\sigma, \ x + k_{0.975}\sigma]$, where $k_{0.975}$ is given by (7.20).

Example 6.1 involved three θ_i quantities but four independent components of error e_a, e_b, e_c and e_d. The parent variances of e_a, e_b and e_c have been specified but the forms of the parent distributions have not. Adding this information helps us illustrate the use of Result 7.2.

Example 6.1 (continued) The distance d between the two points in the plane is measured. The estimate equation is

$$\hat{d} = \sqrt{\hat{r}_1^2 + \hat{r}_2^2 - 2\hat{r}_1\hat{r}_2 \cos\hat{\phi}}.$$

The data are $\hat{r}_1 = 54.8$ m, $\hat{r}_2 = 78.2$ m and $\hat{\phi} = 0.484$ rad and the measurement estimate is $\hat{d} = 39.14$ m. The errors in the estimates of the three unknowns are

$$\hat{r}_1 - r_1 = r_1 e_a + e_b,$$
$$\hat{r}_2 - r_2 = r_2 e_a + e_c,$$
$$\hat{\phi} - \phi = e_d$$

with the four error components e_a, e_b, e_c and e_d having been drawn independently. The measurement error under the linear approximation was given in (7.8), which can also be written as

$$e' = a_a e_a + a_b e_b + a_c e_c + a_d e_d$$

with

$$\begin{aligned}
a_a &= \{(\hat{r}_1 - \hat{r}_2 \cos\hat{\phi})r_1 + (\hat{r}_2 - \hat{r}_1 \cos\hat{\phi})r_2\}/\hat{d} & \approx & \ 39.14, \\
a_b &= (\hat{r}_1 - \hat{r}_2 \cos\hat{\phi})/\hat{d} & = & -0.368, \\
a_c &= (\hat{r}_2 - \hat{r}_1 \cos\hat{\phi})/\hat{d} & = & \ 0.759, \\
a_d &= \hat{r}_1 \hat{r}_2 \sin\hat{\phi}/\hat{d} & = & \ 50.95,
\end{aligned}$$

the figure for a_a being approximate because r_1 and r_2 are approximated by $\hat{r}_1$ and $\hat{r}_2$. This expresses e' as a weighted sum of the realizations of four independent variates.

Suppose that the parent distribution of e_a is isosceles triangular and the parent distributions of e_b and e_c are normal. The error model (6.4)–(6.7) is now

$$e_a \leftarrow T(-0.002 \times \sqrt{6},\ 0.002 \times \sqrt{6}), \tag{7.23}$$
$$e_b \leftarrow N(0, 0.1^2),$$
$$e_c \leftarrow N(0, 0.1^2),$$
$$e_d \leftarrow 0.00408\, t_6,$$

with the last equation being (6.11). So the parent distributions are now fully specified (and the variances are unchanged). From Table 7.2, we see that

$$\sigma_p^2(e_a) = 4 \times 10^{-6}, \qquad \gamma_p(e_a) = -0.6,$$
$$\sigma_p^2(e_b) = 0.01, \qquad \gamma_p(e_b) = 0,$$
$$\sigma_p^2(e_c) = 0.01, \qquad \gamma_p(e_c) = 0,$$
$$\sigma_p^2(e_d) = 2.50 \times 10^{-5}, \qquad \gamma_p(e_d) = 3.$$

Result 7.2 is then applicable. We find that $\sigma_p(e') \approx 0.279$ m and that $\gamma_p(e') \approx 2.065$. So $k_{0.975} \approx 1.998$ and a realized approximate 95% confidence interval for θ is

$$[38.58, \ 38.70] \text{ m}. \tag{7.24}$$

In this case, the parent distribution of the measurement error has a positive coefficient of excess. So the interval is slightly wider than the interval $[38.59, 39.69]$ m calculated using the normal-approximation method.

A Type B error with an uncertain scale

In some situations an experimenter will have in mind a form of parent distribution for a Type B error but will not be sure of the appropriate scale factor. For example a uniform distribution might be thought realistic, but there might be doubt about the width of this distribution. The scale factor might then be regarded as having been drawn from some distribution of its own. The relevant theory is as follows.

Theory 7.2 Let $f_{Y|\Phi=w}(z)$ be the probability density function of a variate Y conditional on the value of a parameter Φ taking the dummy value w, and let $f_\Phi(w)$ be the probability density function of Φ. Then the unconditional probability density function of Y is

$$f_Y(z) = \int_{-\infty}^{\infty} f_{Y|\Phi=w}(z) f_\Phi(w) \, dw,$$

which is a weighted integral of the conditional probability density function.

In our context, w is a scale factor (or scale parameter). We see that attributing a known parent distribution to the scale parameter of an error distribution is, in effect, assigning a known distribution to that error. So it becomes incorrect to describe the overall parent distribution of that error as unknown. Also, in general, the parent distribution for the Type B error is transformed so that its final form is not the one first envisaged.

The unconditional distribution of a variate where an underlying parameter itself has a continuous distribution is called a *compound* distribution. So Y is a compound variable. Our interest, however, is not in the density function of such a variable but in its moments. The theory continues:

7.3 Evaluation using higher moments

Theory 7.2 (continued) The rth moment about the origin of a variate with probability density function $f(z)$ is $\int_{-\infty}^{\infty} z^r f(z)\,dz$, which is linearly related to $f(z)$. Therefore the rth unconditional moment of Y about the origin is given by the same weighted integral of this conditional moment, i.e.

$$\mathcal{E}(Y^r) = \int_{-\infty}^{\infty} \mathcal{E}(Y^r | \Phi = w) f_\Phi(w)\,dw.$$

In the metrological context, (i) the parent mean of the error at any scale factor will almost always be zero and (ii) the parent distribution of the scale factor itself will usually be regarded as uniform. In that case, the following result applies.

Result 7.3 Let e_w be a component of error drawn from a symmetric distribution with mean zero but uncertain scale factor. Let σ^2 and γ be the parent variance and parent coefficient of excess of e_w that would apply if the scale factor were c (such as might be read from Table 7.2) and let the scale factor be regarded as originating from a uniform distribution on the interval $[(1 - \delta)c, (1 + \delta)c]$. Then the parent variance and parent coefficient of excess of e_w unconditional on the scale factor are

$$\sigma_p^2(e_w) = \left(1 + \frac{\delta^2}{3}\right)\sigma^2 \qquad (7.25)$$

and

$$\gamma_p(e_w) = \frac{\left(1 + 2\delta^2 + \delta^4/5\right)\gamma + 4\delta^2 + 4\delta^4/15}{\left(1 + \delta^2/3\right)^2}. \qquad (7.26)$$

(The fact that $\gamma \geq -2$ can then be used to show that $\gamma_p(e_w) \geq \gamma$.)

To see how this result is applicable we will introduce a small modification to Example 6.1.

Example 6.1 (modified) The measurement error in the estimate of distance between the two points is

$$e \approx 39.14\,e_a - 0.368\,e_b + 0.759\,e_c + 50.95\,e_d,$$

with the parent distribution of e_a being the isosceles triangular distribution on the interval $[-0.002 \times \sqrt{6},\ 0.002 \times \sqrt{6}]$, as in (7.23). Suppose now that the experimenter is uncertain about the half-width of the isosceles triangular distribution and that the figure of $0.002 \times \sqrt{6}$ is only accurate to $\pm 50\%$. So $\delta = 0.5$ and the true half-width is known only to lie between $0.001 \times \sqrt{6}$ and $0.003 \times \sqrt{6}$. If the parent distribution of the half-width is uniform between these limits then the model of the generation of e_a is

$$e_a \leftarrow \mathrm{T}(-w, w), \qquad w \leftarrow \mathrm{U}(0.001 \times \sqrt{6},\ 0.003 \times \sqrt{6}) \qquad (7.27)$$

instead of (7.23). From Table 7.2, the variance and coefficient of excess of the isosceles triangular distribution on the condition that the scale factor is $0.002 \times \sqrt{6}$ are seen to be

$$\sigma^2 = 4 \times 10^{-6}, \qquad \gamma = -0.6.$$

Subsequent application of Result 7.3 with $\delta = 0.5$ shows that the unconditional parent variance and parent coefficient of excess of e_a are

$$\sigma_p^2(e_a) = 4.33 \times 10^{-6}, \qquad \gamma_p(e_a) = 0.093.$$

From Result 7.2, the parent variance and parent coefficient of excess of e become

$$\sigma_p^2(e) \approx 0.0786, \qquad \gamma_p(e) \approx 2.043.$$

We obtain $k_{0.975}\sigma_p(e) \approx 0.560$ and find that $[38.58, 38.70]$ m is a realized approximate 95% confidence interval for θ. To the level of precision quoted, this interval is the same as (7.24); the modification affects the limits in the third decimal place. In this situation, the lack of certainty about the appropriate scale factor has made very little difference, even with δ as large as 0.5.

(This example also shows that, when δ is large, the coefficient of excess of the compound distribution can be positive even though the coefficient of excess of the conditional distribution is negative.)

Although Result 7.3 has been presented in this section, (7.25) is also applicable in the normal-approximation method described in Section 7.2.

A Type A error involving five observations or fewer

As described in Section 6.4, the distribution that we suggest be attributed to an error e_i when its parent variance is estimated from the spread of n normal observations is the distribution of $(s/\sqrt{n})T_{n-1}$, with s being the usual sample standard deviation. The parent variance and parent coefficient of e_i are therefore taken to be

$$\sigma_p^2(e_i) = \frac{s^2(n-1)}{n(n-3)}, \qquad n \geq 4,$$

and

$$\gamma_p(e_i) = \frac{6}{n-5}, \qquad n \geq 6.$$

The variance is finite only when $n \geq 4$ and the coefficient of excess is finite only when $n \geq 6$. Therefore, the method only seems applicable when all Type A errors are assessed using six of more observations, and this seems to be a severe restriction. However, it must be remembered that (i) the use of a t distribution as a parent distribution for a Type A error is an approximation, (ii) the use of a Pearson

7.3 Evaluation using higher moments

Table 7.3. *Parent variance and parent coefficient of excess γ assigned to errors of Type A from $n < 6$ observations*

n	Parent variance	γ
5	$0.4\,s^2$	6
4	$0.75\,s^2$	6
3	$1.87\,s^2$	6
2	$202.3\,s^2$	6

distribution is also an approximation and (iii) we are invoking the linear approximation to the measurand equation. So – without changing the status of the method as being approximate – we are at liberty to impute finite parent moments to e_a when the sample size is small. The figures suggested are those in Table 7.3, which supplements Table 7.2 (Willink, 2006b). These figures are obtained as follows.

When $n = 5$ or $n = 4$ the variance of the distribution of T_{n-1} is finite but the coefficient of excess is not. In these cases we use the correct variance and set the excess of the distribution to be the figure for T_5, which is $\gamma = 6$. When $n = 3$ or $n = 2$ we use the variance of the t distribution truncated at the quantiles bounding 99.9% of the probability content (and renormalized) and we set $\gamma = 6$.

Comments on the 'cumulants method'

We shall call the approach described in this section the *cumulants method* (Willink, 2005). (Other possibilities would be the 'method of higher moments' and the 'method of propagating moments'.) The underlying concepts of the method were originally demonstrated in a context where the distributions were attributed not to potential errors but to fixed θ_i quantities, as would be deemed meaningful in a Bayesian analysis. As was explained in Part I, the idea that fixed quantities can be assigned probability distributions is not favoured in this book.

It can be seen that γ, which is the parent coefficient of excess of the linearized error e', is a linear combination of the γ_i (or γ_w) values. It can also be seen that negative values of γ_i (or γ_w) are associated with non-normal Type B errors and that positive values of γ_i (or γ_w) are associated with Type A errors. So – with Type B errors being dominant in many practical measurement situations – we suggest that if the cumulants method is employed then γ will usually be found to be negative. Only if a Type A error is large or if there is a compound distribution formed with a large value of δ might γ exceed zero. Thus, application of the method will typically lead to the legitimate quotation of an interval shorter than that obtained using the normal-approximation method.

The cumulants method can be extended to include asymmetric components of error, asymmetric Pearson distributions and other confidence levels. Equations similar to (7.20) and (7.21) have been obtained for the 0.005, 0.010, 0.025, 0.050, 0.950, 0.975, 0.990 and 0.995 quantiles of standardized asymmetric Pearson distributions (Willink, 2006b).

The method may be criticized for potentially involving an approximating distribution with an unrealistic form. As γ increases from -1.5 to 0 the symmetric Pearson distribution is a unimodal beta distribution, which does not necessarily have a form that is realistic for the sum of error variates. (See the distribution with $\gamma = -1.07$ in Figure 7.1, for example.) A solution is to replace the family of symmetric beta distributions by a family comprising distributions formed as convolutions and containing the uniform and normal distributions as the extremal cases. One possibility is suggested in Appendix F.

Complexity

If θ is subsequently to act as a θ_i quantity in a later measurement then appropriate uncertainty information about θ must be recorded. Both the normal-approximation method of Section 7.2 and the cumulants method of this section require storage of the estimate x and the parent variance attributed to the linearized error e', while the cumulants method also requires storage of the corresponding coefficient of excess γ. The complexity of the cumulants method is therefore similar to that of the method of the *Guide* (BIPM et al., 1995; JCGM, 2008a), which requires storage of x, a figure of standard uncertainty and an accompanying number of 'degrees of freedom'.

Is this level of complexity warranted from a practical point of view? The answer perhaps depends on the sets of problems in which it is applied. If $|\gamma|$ is small in routine problems, so that the parent distribution for the measurement error is close to normal, then the normal-approximation method seems adequate. Furthermore, if the values of $(k_{0.975} - 1.96)$ in routine problems are small and of varying sign then the fluctuations in the probability of success when preferring the simpler procedure will tend to cancel out, in which case this argument for simplicity is strengthened. So the need to involve the coefficient of excess might be questioned in practice. However, against this is the idea that the procedure found in the *Guide* and the procedure found in Supplement 1 to the *Guide* (JCGM, 2008b) both seem based on the idea that the normal-approximation method is potentially inadequate. The cumulants method is a logically sound technique that involves the same premise.

However, there is an important issue that we have overlooked, and this issue has a considerable impact on the complexity of any suitable methodology for uncertainty evaluation. Potentially valuable information is lost when we fail to

7.3 Evaluation using higher moments

separate random and systematic components of the overall error. Accordingly, it is suggested that separate variances and separate coefficients of excess be calculated for these two types of error. Although correct, Result 7.2 becomes inappropriate from the point of view of storing uncertainty information. The following result, which is a reexpression of Result 7.2, is of greater value.

Result 7.4 Let x be an estimate of θ. Suppose that the error $x - \theta$ can be approximated by the sum of components drawn from independent zero-mean symmetric distributions according to the equation

$$x - \theta \approx \sum_r a_r e_r + \sum_s a_s e_s,$$

where 'r' indicates a random error and 's' indicates a systematic error. Let the parent distribution of e_r have variance $\sigma_p^2(e_r)$ and coefficient of excess $\gamma_p(e_r)$, and let the parent distribution of e_s have variance $\sigma_p^2(e_s)$ and coefficient of excess $\gamma_p(e_s)$. Calculate

$$\sigma_{\text{ran}}^2 = \sum_r a_r^2 \sigma_p^2(e_r),$$

$$\sigma_{\text{sys}}^2 = \sum_s a_s^2 \sigma_p^2(e_s),$$

$$\gamma_{\text{ran}} = \frac{\sum_r a_r^4 \gamma_p(e_r) \sigma_p^4(e_r)}{\sigma_{\text{ran}}^4},$$

$$\gamma_{\text{sys}} = \frac{\sum_s a_s^4 \gamma_p(e_s) \sigma_p^4(e_s)}{\sigma_{\text{sys}}^4}$$

and

$$\gamma = \frac{\gamma_{\text{ran}} \sigma_{\text{ran}}^4 + \gamma_{\text{sys}} \sigma_{\text{sys}}^4}{(\sigma_{\text{ran}}^2 + \sigma_{\text{sys}}^2)^2}.$$

If $-1.5 \le \gamma \le 100$ then a realized approximate 95% confidence interval for θ is $[x - k_{0.975}\sigma, x + k_{0.975}\sigma]$, where $k_{0.975}$ is given by (7.20) and

$$\sigma = \sqrt{\sigma_{\text{ran}}^2 + \sigma_{\text{sys}}^2}.$$

So we suggest that when the normal-approximation method is used the information to be stored in order that θ can act as an input to another measurement is x, σ_{ran}^2 and σ_{sys}^2, and when the cumulants method is being used the information required is x, σ_{ran}^2, σ_{sys}^2, γ_{ran} and γ_{sys}. This idea of retaining knowledge of the random and systematic components of error warrants special attention, so it is the principal subject of Chapter 9.

7.4 Monte Carlo evaluation of the error distribution

Sections 7.2 and 7.3 have given two methods for approximating the 0.025 and 0.975 quantiles of the distribution of E' defined in (7.6), which is the error variate applicable under the linear approximation to the measurand equation. A third method is Monte Carlo evaluation, where values drawn from the distributions of $E_1, \ldots, E_m$ are summed according to (7.6) and the distribution of E' is built up by repeating this process many times.

When the E_i errors are dependent, a Monte Carlo analysis would seem to involve sampling from a multivariate distribution, and this could be problematic. However, in most cases any dependence between error variates either (i) arises because the errors are formed from components that themselves are independent or (ii) can be represented in that way. So overall, the m errors are formed from a larger number of independent components and it will only be necessary to sample from univariate distributions. This is the situation in Example 6.1.

Example 6.1 (continued using (7.27)) The distance d between two points is measured. We have seen that the measurement error is $e \approx e'$ with

$$e' \approx 39.14\, e_a - 0.368\, e_b + 0.759\, e_c + 50.95\, e_d,$$

where the errors e_a, e_b, e_c and e_d were incurred independently. This independence makes the simulation of a value of e' straightforward.

The model for these errors is

$$w \leftarrow \mathrm{U}\left(0.001 \times \sqrt{6},\; 0.003 \times \sqrt{6}\right),$$
$$e_a \leftarrow \mathrm{T}(w,\, w),$$
$$e_b \leftarrow \mathrm{N}(0,\, 0.1^2),$$
$$e_c \leftarrow \mathrm{N}(0,\, 0.1^2),$$
$$e_d \leftarrow 0.00408\, \mathrm{t}_6.$$

So the process of generating the linearized error in a measurement can be simulated in the following way.

1. Draw $\tilde{w}$ from the distribution $\mathrm{U}(0.001 \times \sqrt{6},\; 0.003 \times \sqrt{6})$.
2. Draw $\tilde{e}_a$ from the distribution $\mathrm{T}(-\tilde{w},\, \tilde{w})$.
3. Draw $\tilde{e}_b$ from the distribution $\mathrm{N}\left(0,\, 0.1^2\right)$.
4. Draw $\tilde{e}_c$ from the distribution $\mathrm{N}\left(0,\, 0.1^2\right)$.
5. Draw $\tilde{e}_d$ from the distribution $0.00408\, \mathrm{t}_6$.
6. Calculate $\tilde{e}' = 39.14\, \tilde{e}_a - 0.368\, \tilde{e}_b + 0.759\, \tilde{e}_c + 50.95\, \tilde{e}_d$.

(Recall that a tilde ~ is used to indicate a value obtained by simulation.) This process is carried out 10^6 times, and the set of $\tilde{e}'$ values is sorted from lowest to highest. The elements in positions 25 000 and 975 000 are found to be -0.556 and 0.556.

7.4 Monte Carlo evaluation of the error distribution

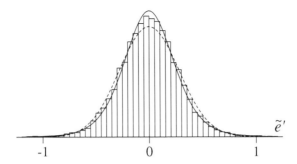

Figure 7.4 Histogram of simulated values of the error under linearization, $\tilde{e}'$. Dashed line: the normal approximation of Section 7.2; solid line: the Pearson approximation of Section 7.3.

These figures will be close approximations to the 0.025 and 0.975 quantiles of the distribution of E'. So an approximate 95% realized confidence interval for d is $[39.141 - 0.556, 39.141 + 0.556]$. This would be quoted as the interval $[38.58, 39.70]$, which is the same as the interval obtained using the cumulants method.

The histogram is shown in Figure 7.4 together with (i) the normal approximation and (ii) the Pearson approximation obtained in the cumulants method. The Pearson approximation is seen to fit the histogram better than the normal distribution.

The number of trials that was used in this example, 10^6, is sufficient to create a histogram that follows the form of the density function closely. However, as we shall now see, this number is much larger than is needed for the calculation of a valid confidence interval.

How many trials?

To answer the question of how many trials are required in such an analysis, we must recall the objective of any method of uncertainty evaluation. We have argued that this objective is *to generate short intervals that enclose the measurand on at least 95% of occasions, with none of these intervals subsequently being seen as more reliable than any other*. There is no need to use Result 6.1 to achieve this objective. Instead we can use Result 6.2, the essence of which is, loosely,

a realized 95% confidence interval for θ is $[x - b, x - a]$, where $[a, b]$ is the output of a procedure that, on average, generates intervals containing 95% of the parent distribution of the error, $x - \theta$.

This fact allows the use of a much smaller number of trials than would typically be involved when locating $e_{0.025}$ and $e_{0.975}$.

The necessary amount of theory is minimal.

Theory 7.3 Let $Y_{(j)}$ be the variate for the jth smallest element in a random sample of size n from some distribution with distribution function F. Then $F\left(Y_{(j)}\right)$ is the variate for the jth smallest element in a random sample of size n from the uniform distribution U(0, 1). This means that

$$\mathcal{E}\left[F\left(Y_{(j)}\right)\right] = \frac{j}{n+1}$$

(David, 1970, Example 3.1.1). That is, the expected value of the area under the probability density function to the left of the jth smallest element in the sample is $j/(n+1)$.

Let $\tilde{e}_{(j)}$ be the jth smallest simulated error in n independent trials. Theory 7.3 implies that, other sources of error notwithstanding, the success rate of our procedure of calculating a confidence interval is 0.95 if we report the interval $[x - \tilde{e}_{(k)}, x - \tilde{e}_{(j)}]$, where

$$\frac{k - j}{n + 1} = 0.95. \tag{7.28}$$

Moreover, we can ensure that the method fails with equal rarity on either side by setting $j + k = n + 1$. The interval calculated in the measurement at hand is then

$$[x - \tilde{e}_{(\overline{0.975\,n+1})},\ x - \tilde{e}_{(\overline{0.025\,n+1})}].$$

So we can put into place a procedure that has the required success rate of 0.95 simply by choosing n, k and j to satisfy (7.28), and n does not have to be large for this to be achieved. Also, all intervals calculated using this method can be subsequently regarded as equally reliable. Therefore – according to our criteria for a good procedure – any advantage of using a large number of trials must be related to the shortness of the intervals produced. The limited extent of this advantage is illustrated in the following analysis.

Suppose that a quantity θ is measured and the result is x. The error is $e = x - \theta$. Consider evaluating the measurement uncertainty by applying the Monte Carlo procedure with n trials, with $(k - j)/(n + 1) = 0.95$ and with $k + j = n + 1$. If n is infinite then the procedure generates the interval $[x - e_{0.975}, x - e_{0.025}]$. If the parent distribution of e is N(0, 1) then this interval is $[x - 1.96, x + 1.96]$, which has length 3.92. If the parent distribution is U(-1, 1) then this interval is $[x - 0.95, x + 0.95]$, which has length 1.9. And if the parent distribution is A(-5, 5) then this interval is $[x - 4.9846, x + 4.9846]$, which has length 9.9692. The procedure was applied 10^5 times for each of these three distributions and for each of several values of n.

7.4 Monte Carlo evaluation of the error distribution

Table 7.4. *Means and standard deviations of the widths of realized 95% confidence intervals obtained using n trials*

			N(0, 1)		U(−1, 1)		A(−5, 5)	
n	j	k	$\bar{w}$	σ_w	$\bar{w}$	σ_w	$\bar{w}$	σ_w
39	1	39	4.302	0.669	1.900	0.068	9.941	0.087
79	2	78	4.106	0.446	1.900	0.048	9.954	0.047
199	5	195	3.991	0.272	1.900	0.031	9.963	0.023
999	25	975	3.934	0.119	1.900	0.014	9.968	0.009
9999	250	9750	3.921	0.037	1.900	0.004	9.969	0.003
∞			3.920		1.900		9.969	

In every case the success rate observed was consistent with the theoretical figure, 0.95. The mean lengths of the intervals are seen in Table 7.4. Increasing n reduced the mean length of intervals when the error distribution was normal, had no effect when it was uniform and actually increased the mean length when it was the arcsine distribution! (These results may seem reasonable after some reflection.) The table also shows the standard deviations of the lengths of the intervals. The same analysis was carried out for 99% uncertainty intervals, which are found by setting $(k - j)/(n + 1) = 0.99$. The same behaviour was observed.

We infer that increasing the number of trials in this type of Monte Carlo procedure will reduce the mean length of the intervals generated when the error distribution has a bell-shape of some form, as in the typical measurement problem. However, the potential reduction in length is small above moderate values of n. When the distribution is normal, choosing $n = 9999$ is sufficient to obtain 95% and 99% intervals with mean lengths only approximately 0.04% and 0.1% greater than the minimum values. (With $n = 999$, the respective figures are 0.4% and 1%.) Therefore, we suggest that 9999 is a suitable number of trials. Performance will be slightly worse when the distribution has considerable kurtosis, such as a t distribution with few degrees of freedom. Simulations show that in the extreme case of a single degree of freedom the mean lengths of 95% and 99% intervals with $n = 9999$ are approximately 25.51 and 129.9, which are approximately 0.4% and 2% greater than the lengths 25.41 and 127.3 that correspond to infinite n.

Let us see how the recommended value of $n = 9999$ performs in our example of distance measurement.

> **Example 6.1 (continued using (7.27))** The distance d between two points is measured. The simulated linearized error in a measurement is found using steps 1–6 in the description given earlier in this section. This process is carried out $n = 9999$ times, and the set of $\tilde{e}'$ values is sorted from lowest to highest. The elements in positions 250 and 9750 of this ordered set are found to be −0.557 and 0.567. So an

approximate realized 95% confidence interval for d is [39.141 − 0.567, 39.141 + 0.557], which has length 1.124. This interval might be quoted as the interval [38.57, 39.70], which happens to be slightly longer than the interval [38.58, 39.70] obtained with 10^6 trials.

This whole procedure was carried out 1000 times in order to assess the distribution of lengths of the intervals generated with $n = 9999$. The sample of 1000 lengths had mean 1.111, median 1.111 and standard deviation 0.013. The lengths in the sample ranged from 1.070 to 1.149.

Using a finite value of n leads to different intervals on different occasions, but that does not seem important. Some readers might disagree and might insist that the quantiles $e_{0.025}$ and $e_{0.975}$ should always be located with high accuracy, and that n should be large enough to achieve this. However, to hold to the idea that there is a unique 'correct' uncertainty interval to be calculated in any one situation is to fail to acknowledge that the practical meaning of probability is only found in long-run performance, as was shown in Section 3.1: the idea of attributing a probability to an event only has meaning in some long-run context.

Another argument can also be made for accepting variability in the output of the Monte Carlo analysis. Consider a measurement with an influential Type A error involving a sample mean x_i and sample variance s_i^2. Repetition of the procedure would lead to different values of x_i and s_i^2, and so to uncertainty intervals of different length, (as well as different measurement estimates). If we accept such variability in the intervals obtained when there is Type A evaluation of uncertainty then why should we not permit variability when using a Monte Carlo analysis?

In summary, let us make two points. First, the reason for using a large number of trials in a Monte Carlo analysis of this kind is related to the length of the intervals, not to the achievement of the advertised minimum success rate of 95%. Second, in the typical problem the mean interval length obtained with $n = 9999$ will only exceed the length with n infinite by an amount of the order of 0.1%. So it is suggested that choosing $n = 9999$ is adequate.

Using symmetry

In almost every case, the distribution of each E_i variate will be symmetric and have mean zero, so the distribution of E' will also be symmetric and have mean zero. Then the quantities $-e'_{0.025}$ and $e'_{0.975}$ are both equal to the 0.95 quantile of the distribution of $|E'|$, and we can work with the ordered set of $|\tilde{e}'|$ values instead. Subsequently Result 6.2 and Theory 7.3 together imply that a realized approximate 95% confidence interval for θ is $[x − q, x + q]$, where q is the 0.95 quantile of this set.

8
Evaluation without the linear approximation

Usually the distribution of E will be well approximated by the distribution of E', so that the linear approximation will be adequate. However, sometimes the structure of the measurand equation $\theta = \mathcal{F}(\theta_1, \ldots, \theta_m)$, the extent of its non-linearity and the sizes of the standard deviations $\{\sigma(E_1), \ldots, \sigma(E_m)\}$ will mean that this is not the case. In this chapter, we discuss how an interval of measurement uncertainty can be derived in such a situation.

Section 8.1 discusses the use of higher-order terms in the Taylor expansion and shows that, in general, the measurement result $x = \mathcal{F}(x_1, \ldots, x_m)$ is a biased estimate of the target value θ. Section 8.2 shows how quantities with estimates that vary with repetition of the measurement and quantities with fixed estimates require different treatment when non-linearity is taken into account. Sections 8.3–8.5 consider the use of Monte Carlo methodology to simulate the measurement and to put statistical bounds on the values of the θ_i quantities that could have given rise to the data. This enables the calculation of a confidence interval for θ. In the textbook situation this will be an ordinary confidence interval, but in practice it might be a conditional confidence interval or an average confidence interval.

8.1 Including higher-order terms

The methods of analysis discussed in Chapter 7 were based on the linear approximation to the measurand equation. We now consider an analysis that also uses higher-order terms of the Taylor's series.

In the textbook situation where every x_i estimate takes a different value when the measurement is repeated, x_i is seen as the outcome of a variate $\theta_i + E_i$ and x is seen as the outcome of a variate

$$X = \mathcal{F}(\theta_1 + E_1, \ldots, \theta_m + E_m).$$

From this, and with $\boldsymbol{\theta} = (\theta_1, \ldots, \theta_m)$, we can write

$$X = \theta + \sum_{i=1}^{m} E_i \frac{\partial \mathcal{F}(\boldsymbol{\theta})}{\partial \theta_i} + \sum_{i=1}^{m} \sum_{j=1}^{m} \frac{E_i E_j}{2} \frac{\partial^2 \mathcal{F}(\boldsymbol{\theta})}{\partial \theta_i \partial \theta_j}$$

$$+ \sum_{i=1}^{m} \sum_{j=1}^{m} \sum_{k=1}^{m} \frac{E_i E_j E_k}{6} \frac{\partial^3 \mathcal{F}(\boldsymbol{\theta})}{\partial \theta_i \partial \theta_j \partial \theta_k}$$

$$+ \text{ terms of order four or more.} \tag{8.1}$$

Consider the measurement-error variate $E = X - \theta$ when each E_i is independent and has mean zero. Taking the mean of each term on the right-hand side of (8.1) gives

$$\mathcal{E}(E) = \sum_{i=1}^{m} \frac{\sigma^2(E_i)}{2} \frac{\partial^2 \mathcal{F}(\boldsymbol{\theta})}{\partial \theta_i^2} + \text{ terms of order four or more.} \tag{8.2}$$

Thus, X estimates θ with a bias that depends on the second partial derivatives of $\mathcal{F}(\boldsymbol{\theta})$, and so an estimate of θ that is partially corrected for bias is

$$x^* = x - \sum_{i=1}^{m} \frac{\sigma^2(E_i)}{2} \frac{\partial^2 \mathcal{F}(\mathbf{x})}{\partial x_i^2} \tag{8.3}$$

with $\mathbf{x} = (x_1, \ldots, x_m)$. This estimate might be preferred to x.

To apply a similar adjustment to the uncertainty statement we recognize that

$$\text{var} E = \mathcal{E}(E^2) - \mathcal{E}(E)^2 \tag{8.4}$$

and then write E in the form

$$E = \sum_i E_i \frac{\partial \mathcal{F}}{\partial \theta_i}$$

$$+ \sum_i \frac{E_i^2}{2} \frac{\partial^2 \mathcal{F}}{\partial \theta_i^2} + \sum_{i<j} E_i E_j \frac{\partial^2 \mathcal{F}}{\partial \theta_i \partial \theta_j}$$

$$+ \sum_i \frac{E_i^3}{6} \frac{\partial^3 \mathcal{F}}{\partial \theta_i^3} + \sum_{i \neq j} \frac{E_i E_j^2}{2} \frac{\partial^3 \mathcal{F}}{\partial \theta_i \partial \theta_j^2} + \sum_{i<j<k} E_i E_j E_k \frac{\partial^3 \mathcal{F}}{\partial \theta_i \partial \theta_j \partial \theta_k}$$

$$+ \text{ terms of order four or more,}$$

with $\mathcal{F}$ without argument indicating $\mathcal{F}(\boldsymbol{\theta})$. Because of the independence of the E_i variates, a term in which any E_i appears to the power of 1 has zero expectation. We also suppose that these variates are symmetric, which means that a term in which any E_i appears to any other positive odd power also has zero expectation. So, with σ_i^2 indicating $\sigma^2(E_i)$, the expected value of E^2 is

8.1 Including higher-order terms

$$\mathcal{E}(E^2) = \sum_i \sigma_i^2 \left(\frac{\partial \mathcal{F}}{\partial \theta_i}\right)^2 + \sum_i \frac{\mathcal{E}E_i^4}{4}\left(\frac{\partial^2 \mathcal{F}}{\partial \theta_i^2}\right)^2 + \sum_i \sum_{i \neq j} \frac{\sigma_i^2 \sigma_j^2}{4} \frac{\partial^2 \mathcal{F}}{\partial \theta_i^2} \frac{\partial^2 \mathcal{F}}{\partial \theta_j^2}$$

$$+ \sum_i \sum_{i \neq j} \frac{\sigma_i^2 \sigma_j^2}{2}\left(\frac{\partial^2 \mathcal{F}}{\partial \theta_i \partial \theta_j}\right)^2 + \sum_i \frac{\mathcal{E}E_i^4}{6} \frac{\partial \mathcal{F}}{\partial \theta_i} \frac{\partial^3 \mathcal{F}}{\partial \theta_i^3}$$

$$+ \sum_i \sum_{i \neq j} \frac{\sigma_i^2 \sigma_j^2}{2} \frac{\partial \mathcal{F}}{\partial \theta_i} \frac{\partial^3 \mathcal{F}}{\partial \theta_i \partial \theta_j^2}$$

+ terms of order six or more. (8.5)

The complexity of this expression indicates that a considerable amount of work is needed to include all fourth-order terms in the evaluation of uncertainty, even in the independent case.

The effects of including the non-linear terms are now examined in a simple situation where many of the partial derivatives vanish. The results are surprising.

Example 8.1 Suppose that the measurand is a total power, with the target value being representable by $\theta = \mathcal{F}(\theta_1, \theta_2) = \theta_1^2 + \theta_2^2$. The estimator of θ is $X = X_1^2 + X_2^2$ and the estimate is $x = x_1^2 + x_2^2$. In this situation

$$\frac{\partial \mathcal{F}}{\partial \theta_i} = 2\theta_i, \qquad \frac{\partial \mathcal{F}^2}{\partial \theta_i^2} = 2, \qquad i = 1, 2$$

and

$$\frac{\partial \mathcal{F}(\mathbf{x})}{\partial x_i} = 2x_i, \qquad \frac{\partial \mathcal{F}^2(\mathbf{x})}{\partial x_i^2} = 2, \qquad i = 1, 2$$

and all other partial derivatives are zero.

Suppose the errors $E_1 = X_1 - \theta_1$ and $E_2 = X_2 - \theta_2$ are independently drawn from some distribution with mean zero and variance σ^2. Then, from (8.3), $x^* = x - 2\sigma^2$. Suppose also that this distribution is symmetric and has coefficient of excess γ. The last three terms shown in (8.5) are zero because the partial derivatives of higher order are zero, so

$$\mathcal{E}(E^2) = \sum_{i=1}^2 \sigma^2 \left(\frac{\partial \mathcal{F}}{\partial \theta_i}\right)^2 + \sum_{i=1}^2 \frac{(3+\gamma)\sigma^4}{4}\left(\frac{\partial^2 \mathcal{F}}{\partial \theta_i^2}\right)^2 + \frac{\sigma^4}{2} \frac{\partial^2 \mathcal{F}}{\partial \theta_1^2} \frac{\partial^2 \mathcal{F}}{\partial \theta_2^2},$$

which together with (8.4) gives

$$\mathrm{var}\, E = \left\{4\theta\sigma^2 + 2(3+\gamma)\sigma^4 + 2\sigma^4\right\} - (2\sigma^2)^2.$$

Last, suppose the distribution is normal, so $\gamma = 0$. Then

$$\mathrm{var}\, E = 4\sigma^2(\theta + \sigma^2). \tag{8.6}$$

128   *Evaluation without the linear approximation*

This exact expression may be contrasted with the result that would be obtained under the linear approximation, which is

$$\text{var} E \approx 4\sigma^2 \theta. \tag{8.7}$$

Suppose we replace the unknown value θ by the original estimate x when estimating var E using (8.6) or (8.7). The 95% uncertainty interval for θ calculated using the linear approximation and the original estimate will be

$$\left[x - 3.92\sigma \sqrt{x}, \; x + 3.92\sigma \sqrt{x} \right], \tag{8.8}$$

the interval calculated using the linear approximation and the bias-adjusted estimate x^* will be

$$\left[x^* - 3.92\sigma \sqrt{x}, \; x^* + 3.92\sigma \sqrt{x} \right] \tag{8.9}$$

and the interval calculated using (8.6), the bias-adjusted estimate and the normal approximation will be

$$\left[x^* - 3.92\sigma \sqrt{x + \sigma^2}, \; x^* + 3.92\sigma \sqrt{x + \sigma^2} \right]. \tag{8.10}$$

In contrast, suppose we instead replace the unknown value θ by the bias-adjusted estimate x^* when forming our estimate of var E. The three corresponding intervals are

$$\left[x - 3.92\sigma \sqrt{x^*}, \; x + 3.92\sigma \sqrt{x^*} \right], \tag{8.11}$$

$$\left[x^* - 3.92\sigma \sqrt{x^*}, \; x^* + 3.92\sigma \sqrt{x^*} \right] \tag{8.12}$$

and

$$\left[x^* - 3.92\sigma \sqrt{x^* + \sigma^2}, \; x^* + 3.92\sigma \sqrt{x^* + \sigma^2} \right]. \tag{8.13}$$

Simulations were carried out with $\theta_1 = \theta_2 = 10$ and with different values of σ to examine the accuracy of the adjustment for bias and to compare the success rates when using these six intervals. There were 5×10^6 replications at each value of σ. The results are shown in Table 8.1. In this example, the adjustment for bias is exact. However, as can be seen from the success rates, such an adjustment should

Table 8.1. *Mean estimates and success rates with use of different terms*

σ	$\bar{x}$	$\bar{x}^*$	(8.8)	(8.9)	(8.10)	(8.11)	(8.12)	(8.13)
0.01	200.0	200.0	0.950	0.950	0.950	0.950	0.950	0.950
0.1	200.0	200.0	0.950	0.950	0.950	0.950	0.950	0.950
0.5	200.5	200.0	0.950	0.949	0.950	0.950	0.949	0.949
1	202.0	200.0	0.949	0.947	0.947	0.948	0.946	0.946
2	208.0	200.0	0.947	0.938	0.941	0.941	0.932	0.935
3	218.0	200.0	0.943	0.926	0.934	0.928	0.909	0.918

8.1 Including higher-order terms

not necessarily be applied when forming an uncertainty interval. The best results are actually obtained with the basic interval (8.8), which involves centring the interval on the original estimate x, employing the linear approximation and using the original estimate x in the estimation of var E.

So applying the adjustment for bias in the uncertainty analysis is not necessarily helpful, and seeking to take non-linear terms into account in the variance estimation might be disadvantageous. One cause for the failure of these analytical approaches with non-linear functions is the asymmetry of the distribution of the total error. A second reason is that the methodology is implemented using partial derivatives evaluated at the point $(x_1, \ldots, x_m)$ instead of at the unknown point $(\theta_1, \ldots, \theta_m)$. Any assumption that this step does not introduce unacceptable error in the non-linear case would appear to be faulty.

The simple multiplication of two quantities provides another example of complications that arise when we seek to go beyond the linear approximation.

Example 8.2 Suppose we wish to measure the quantity $\theta = \theta_1 \theta_2$. The estimator is $X = X_1 X_2$ with $X_i = \theta_i + E_i$, so the error variate $E = X - \theta$ is

$$E = E_1 \theta_2 + E_2 \theta_1 + E_1 E_2.$$

Suppose X_i is an unbiased estimator of θ_i, i.e. suppose $\mathcal{E}(E_i) = 0$. Let us write σ_i^2 for $\sigma^2(E_i)$. Then

$$\text{var } E = \theta_2^2 \sigma_1^2 + \theta_1^2 \sigma_2^2 + \sigma_1^2 \sigma_2^2. \tag{8.14}$$

We could replace θ_i by its unbiased estimate x_i in (8.14) to obtain the figure

$$x_2^2 \sigma_1^2 + x_1^2 \sigma_2^2 + \sigma_1^2 \sigma_2^2 \tag{8.15}$$

as one estimate of the variance of E. However, it would perhaps be better to replace θ_i^2 by its unbiased estimate instead, which is $x_i^2 - \sigma_i^2$. (Of course, this figure would not be used if it were negative.) Our estimate of the variance of E then becomes

$$\widehat{\text{var }} E = x_2^2 \sigma_1^2 + x_1^2 \sigma_2^2 - \sigma_1^2 \sigma_2^2, \tag{8.16}$$

which differs from (8.15) in the sign of the term $\sigma_1^2 \sigma_2^2$. The estimates given in (8.15) and (8.16) can be contrasted with the estimate obtained with the basic linear approximation, which is $x_2^2 \sigma_1^2 + x_1^2 \sigma_2^2$.

The only non-zero partial derivatives are

$$\frac{\partial \mathcal{F}}{\partial \theta_1} = \theta_2, \qquad \frac{\partial \mathcal{F}}{\partial \theta_2} = \theta_1, \qquad \frac{\partial \mathcal{F}^2}{\partial \theta_1 \partial \theta_2} = 1.$$

So the adjustment for bias is zero, and the term $+\sigma_1^2 \sigma_2^2$ in (8.15) arises from the contributions to (8.5) that involve $\partial \mathcal{F}^2/(\partial \theta_1 \partial \theta_2)$. However, we see that including that term actually moves the estimate of the error variance further from the figure

8.2 The influence of fixed estimates

In Example 8.1 the ratio of the standard deviation of each individual estimator X_i to its mean θ_i (which is a ratio called the coefficient of variation) was as great as $3/10 = 30\%$, which is a very high figure. When there are such large relative deviations from an operating point $(\theta_1, \ldots, \theta_m)$, we might expect non-linear terms to be important. However, this was not observed in our example. Instead our best results were obtained with the linear approximation. This observation raises the question of when, if ever, the linear approximation is inadequate or can be improved upon in a meaningful way.

We suggest that the principal kind of measurement for which the linear approximation will not be adequate is where a partial derivative evaluated at the vector of measurement estimates $(x_1, \ldots, x_m)$ is zero, so that the parent variance of the corresponding error has no influence at all on the stated uncertainty unless some higher-order terms are included. Usually, a partial derivative will only evaluate to zero when one or more of the estimates has a fixed 'round' value, the usual value being zero. So it is appropriate in our study of non-linearity to consider the proper treatment of situations in which one or more of the x_i values is fixed. As has already been explained, if x_i is fixed then the figure 0.95 quoted with an uncertainty interval will refer to the success rate over a set of measurements in which the quantity with value θ_i is distributed around x_i. The figure of standard uncertainty associated with x_i will be the standard deviation of this distribution.

Let us consider a modified version of Example 8.2, where $\theta = \theta_1\theta_2$. This example might seem unrealistically simple. However, it is relevant to an automated algorithmic method of uncertainty propagation devised by Hall (2006), which is implemented by the decomposition of the measurand equation into successive steps involving binary operations such as addition and multiplication.

Example 8.2 (modified) We wish to measure the quantity $\theta = \theta_1\theta_2$. Our estimates of θ_1 and θ_2 are x_1 and x_2, and our estimate of θ is $x = x_1x_2$. According to the basic linear approximation, the parent variance of the error $e = x - \theta$ is var$(E) \approx x_2^2\sigma_1^2 + x_1^2\sigma_2^2$. We wish to take the non-linearity of the function into account. Earlier we saw how this might be done when θ_1 and θ_2 were fixed. Now we suppose that one or both of the x_i values is fixed.

Consider the situation where both x_1 and x_2 are fixed. The unknowns θ_1 and θ_2 are then seen as outcomes of variates Θ_1 and Θ_2 that obey

8.2 The influence of fixed estimates

$$\Theta_1 \sim D(x_1, \sigma_1^2),$$
$$\Theta_2 \sim D(x_2, \sigma_2^2),$$

where σ_1^2 and σ_2^2 are the parent variances attributed to the Type B errors $x_1 - \theta_1$ and $x_2 - \theta_2$. The measurement-error variate is

$$E = x_1 x_2 - \Theta_1 \Theta_2.$$

We find that

$$\text{var } E = x_1^2 \sigma_2^2 + x_2^2 \sigma_1^2 + \sigma_1^2 \sigma_2^2. \tag{8.17}$$

Now consider the situation where only x_1 is fixed, so that x_2 is seen as the outcome of a variate X_2 with mean θ_2 and standard deviation σ_2. Here the only distribution of unknown quantities that we invoke is the distribution of Θ_1. The statistical model is now

$$\Theta_1 \sim D(x_1, \sigma_1^2),$$
$$X_2 \sim D(\theta_2, \sigma_2^2).$$

Now $X = x_1 X_2$, $\Theta = \Theta_1 \theta_2$ and $E = X - \Theta$, so

$$\text{var } E = \text{var}(x_1 X_2) + \text{var}(\Theta_1 \theta_2)$$
$$= x_1^2 \sigma_2^2 + \theta_2^2 \sigma_1^2.$$

Our unbiased estimate of θ_2^2 is $x_2^2 - \sigma_2^2$, so our best estimate of var E is

$$\widehat{\text{var}} \, E = x_1^2 \sigma_2^2 + x_2^2 \sigma_1^2 - \sigma_2^2 \sigma_1^2, \tag{8.18}$$

which is the result we obtained in (8.16).

Equations (8.17) and (8.18) differ in the sign of the term $\sigma_1^2 \sigma_2^2$, which is a term that is present only because we are trying to improve on the basic linear approximation. This shows that, when considering non-linear terms, different treatments are required for an estimate that varies from measurement to measurement and an estimate that is fixed.

The matter of bias is also relevant here. The estimator X is negatively biased when both x_1 and x_2 are fixed, but it is unbiased when only x_1 is fixed. When neither estimate is fixed, as was the case in the version of this example given in Section 8.1, the estimator is positively biased.

Let us now consider the general situation of $\theta = \mathcal{F}(\theta_1, \ldots, \theta_m)$. When x_i is fixed we see θ_i as being the outcome of a variate Θ_i with variance σ_i^2. Without losing generality, we suppose the fixed estimates are $x_{k+1}, \ldots, x_m$. Then the measurement error variate is

$$E = \mathcal{F}(X_1, \ldots, X_k, x_{k+1}, \ldots, x_m) - \mathcal{F}(\theta_1, \ldots, \theta_k, \Theta_{k+1}, \ldots, \Theta_m)$$

and the variance of E is

$$\operatorname{var} E = \sum_{i=1}^{k} \frac{\partial \mathcal{F}(\cdot)}{\partial \theta_i} \sigma_i^2 + \sum_{i=k+1}^{m} \frac{\partial \mathcal{F}(\cdot)}{\partial x_i} \sigma_i^2 + \text{terms of higher order,} \qquad (8.19)$$

with

$$\frac{\partial \mathcal{F}(\cdot)}{\partial \theta_i} \equiv \left. \frac{\partial \mathcal{F}(\theta_1, \ldots, \theta_{i-1}, z, \theta_{i+1}, \ldots, \theta_k, x_{k+1}, \ldots, x_m)}{\partial z} \right|_{z=\theta_i}$$

and

$$\frac{\partial \mathcal{F}(\cdot)}{\partial x_i} \equiv \left. \frac{\partial \mathcal{F}(\theta_1, \ldots, \theta_k, x_{k+1}, \ldots, x_{i-1}, z, x_{i+1}, \ldots, x_m)}{\partial z} \right|_{z=x_i}. \qquad (8.20)$$

In general, the linear approximation will be adequate unless the estimate of one or more of the partial derivatives is small. This may well be the case when there is a matching error, in which case the estimate of the quantity is zero. Consider the following example, which is based on one found in Supplement 1 to the *Guide* (JCGM, 2008b, Example 9.3).

Example 8.3 A weight of density ρ and unknown mass m is calibrated against a reference weight of density ρ_{ref} and mass m_{ref} using a balance operated in air of density ρ_{air}. The masses are nominally equal to 50 g, this being the stated fixed estimate of m_{ref}. The densities ρ and ρ_{ref} are also nominally equal, so we write $\rho = \rho_{\text{ref}} + \epsilon$ with ϵ having an estimate of zero. Using Archimedes' principle, we find that the measurand equation is

$$m = (s + m_{\text{ref}}) \frac{1 - \rho_{\text{air}}/\rho_{\text{ref}}}{1 - \rho_{\text{air}}/(\rho_{\text{ref}} + \epsilon)}$$

with s being the mass of a small weight of density ρ_{ref} added to the reference weight to give a balance. The estimate equation is

$$\hat{m} = (\hat{s} + \hat{m}_{\text{ref}}) \frac{1 - \hat{\rho}_{\text{air}}/\hat{\rho}_{\text{ref}}}{1 - \hat{\rho}_{\text{air}}/(\hat{\rho}_{\text{ref}} + \hat{\epsilon})}.$$

Because $\hat{\epsilon} = 0$, the estimates of $\partial \hat{m}/\partial \rho_{\text{air}}$ and $\partial \hat{m}/\partial \rho_{\text{ref}}$ calculated by the usual method are both equal to zero. So unless we include higher-order terms in the analysis (or unless we use other estimates of $\partial \hat{m}/\partial \rho_{\text{air}}$ and $\partial \hat{m}/\partial \rho_{\text{ref}}$) the parent variances of $\hat{\rho}_{\text{air}}$ and $\hat{\rho}_{\text{ref}}$ will not feature at all when assessing the size of the possible error in $\hat{m}$.

One way forward in such a situation would be the development of software featuring the automatic inclusion of all fourth-order terms. In the absence of such software, a practical approach would involve the judicious inclusion of higher-order terms. The starting point might be the premise that the linear approximation is adequate when all of the first-order derivatives have non-zero estimates. Working

8.2 The influence of fixed estimates

on this premise, we suggest that the only higher-order terms to be included are those that involve an error E_i for which the estimate of the corresponding derivative $\partial \mathcal{F}(\cdot)/\partial \theta_i$ or $\partial \mathcal{F}(\cdot)/\partial x_i$ is zero. For each such i, the estimates of the higher-order derivatives $\partial^h \mathcal{F}(\cdot)/\partial \theta_i^h$ or $\partial^h \mathcal{F}(\cdot)/\partial x_i^h$ might well be zero also. In this way, we will find that only a minority of the non-linear terms in (8.5) need to be retained.

Example 8.3 (continued) The measurand equation is

$$\theta = \mathcal{F}(\theta_1, \theta_2, \theta_3, \theta_4, \theta_5) = (\theta_1 + \theta_4)\frac{1 - \theta_2/\theta_3}{1 - \theta_2/(\theta_3 + \theta_5)},$$

with $\theta = m$ and $(\theta_1, \theta_2, \theta_3, \theta_4, \theta_5) = (s, \rho_{air}, \rho_{ref}, m_{ref}, \epsilon)$.

Equations (8.2) and (8.5) are no longer strictly appropriate because, with some of the estimates being fixed, some of the partial derivatives should be taken about these estimates, as in (8.20). However, there will be equations analogous to (8.2) and (8.5) that involve this modification. The estimates x_4 and x_5 are fixed, so the equation analogous to (8.2) is

$$\mathcal{E}(E) \approx \sum_{i=1}^{3} \frac{\sigma_i^2}{2}\frac{\partial^2 \mathcal{F}(\cdot)}{\partial \theta_i^2} + \sum_{i=4}^{5} \frac{\sigma_i^2}{2}\frac{\partial^2 \mathcal{F}(\cdot)}{\partial x_i^2}. \quad (8.21)$$

The estimate of θ_5 is $x_5 = 0$, and consequently both $\partial^h \mathcal{F}(\cdot)/\partial \theta_2^h = 0$ and $\partial^h \mathcal{F}(\cdot)/\partial \theta_3^h = 0$ for $h = 1, 2, 3$. So the only fourth-order terms that we retain in the equation analogous to (8.5) are those involving moments of E_2 or E_3. Of these terms, those that have $\partial^h \mathcal{F}(\cdot)/\partial \theta_2^h$ or $\partial^h \mathcal{F}(\cdot)/\partial \theta_3^h$ as a factor are zero. We are left with the equation

$$\mathcal{E}(E^2) \approx \sum_{i=1}^{3} \sigma_i^2 \left(\frac{\partial \mathcal{F}(\cdot)}{\partial \theta_i}\right)^2 + \sum_{i=4}^{5} \sigma_i^2 \left(\frac{\partial \mathcal{F}(\cdot)}{\partial x_i}\right)^2$$

$$+ \sum_{j=2,3} \sigma_5^2 \sigma_j^2 \left(\frac{\partial^2 \mathcal{F}(\cdot)}{\partial x_5 \partial \theta_j}\right)^2 + \sum_{j=2,3} \frac{\sigma_5^2 \sigma_j^2}{2} \frac{\partial \mathcal{F}(\cdot)}{\partial x_5} \frac{\partial^3 \mathcal{F}(\cdot)}{\partial x_5 \partial \theta_j^2}, \quad (8.22)$$

in which the first two sums represent the linear approximation.

The expectations $\mathcal{E}(E)$ and $\mathcal{E}(E^2)$ can be approximated and the results used with the normal approximation to obtain an uncertainty interval for the target value.

Example 8.3 (continued) The experiment involves obtaining the estimate of s. It is known that such an experiment incurs normally distributed error with mean zero and standard deviation 0.1×10^{-3} g. The estimate of s obtained is 3.4×10^{-3} g. Therefore, the model for the experimental part of the measurement is (in grams)

$$\hat{s} = 3.4 \times 10^{-3},$$
$$\hat{s} - s \leftarrow N\left(0, (0.1 \times 10^{-3})^2\right).$$

The estimates of ρ_{air} and ρ_{ref}, which were obtained in the background part of the measurement, are

$$\hat{\rho}_{\text{air}} = 1.2 \times 10^3 \text{ g m}^{-3},$$
$$\hat{\rho}_{\text{ref}} = 8.1 \times 10^6 \text{ g m}^{-3}$$

and the corresponding error models are (in g m^{-3})

$$\hat{\rho}_{\text{air}} - \rho_{\text{air}} \leftarrow \text{U}(-0.2 \times 10^3, \ 0.2 \times 10^3),$$
$$\hat{\rho}_{\text{ref}} - \rho_{\text{ref}} \leftarrow \text{U}(-0.5 \times 10^6, \ 0.5 \times 10^6).$$

The estimates of $\hat{m}_{\text{ref}}$ and ϵ are the fixed values

$$\hat{m}_{\text{ref}} = 50 \text{ g},$$
$$\hat{\epsilon} = 0 \text{ g m}^{-3}$$

and the corresponding error models are

$$\hat{m}_{\text{ref}} - m_{\text{ref}} \leftarrow \text{N}\left(0, (0.02 \times 10^{-3})^2\right),$$
$$\hat{\epsilon} - \epsilon \leftarrow \text{U}(-1.5 \times 10^6, \ 1.5 \times 10^6).$$

Thus

$$\begin{aligned}
x_1 &= 3.8 \times 10^{-3} \text{ g}, & \sigma_1 &= 0.1 \times 10^{-3} \text{ g}, \\
x_2 &= 1.2 \times 10^3 \text{ g m}^{-3}, & \sigma_2 &= 0.3 \times 10^3 / \sqrt{3} \text{ g m}^{-3}, \\
x_3 &= 8.2 \times 10^6 \text{ g m}^{-3}, & \sigma_3 &= 0.5 \times 10^6 / \sqrt{3} \text{ g m}^{-3}, \\
x_4 &= 50 \text{ g}, & \sigma_4 &= 5 \times 10^{-6} \text{ g}, \\
x_5 &= 0 \text{ g m}^{-3}, & \sigma_5 &= 1.5 \times 10^6 / \sqrt{3} \text{ g m}^{-3}.
\end{aligned}$$

(To obtain results that are illustrative, it has been necessary to use some figures of standard deviation that are unrealistic.)

The estimates of the quantities $\mathcal{E}(E)$ and $\mathcal{E}(E^2)$ in (8.21) and (8.22) are obtained using the equations

$$\hat{\mathcal{E}}(E) = \sum_{i=1}^{5} \frac{\sigma_i^2}{2} \frac{\partial^2 \mathcal{F}(\mathbf{x})}{\partial x_i^2} \tag{8.23}$$

and

$$\hat{\mathcal{E}}(E^2) = \sum_{i=1}^{5} \sigma_i^2 \left(\frac{\partial \mathcal{F}(\mathbf{x})}{\partial x_i}\right)^2$$
$$+ \sum_{j=2}^{3} \sigma_5^2 \sigma_j^2 \left(\frac{\partial^2 \mathcal{F}(\mathbf{x})}{\partial x_5 \partial x_j}\right)^2 + \sum_{j=2}^{3} \frac{\sigma_5^2 \sigma_j^2}{2} \frac{\partial \mathcal{F}(\mathbf{x})}{\partial x_5} \frac{\partial^3 \mathcal{F}(\mathbf{x})}{\partial x_5 \partial x_j^2}. \tag{8.24}$$

The partial derivatives in (8.23) and (8.24) are evaluated numerically, and we obtain

$$\hat{\mathcal{E}}(E) \approx 82 \times 10^{-6} \text{ g},$$
$$\hat{\mathcal{E}}(E^2) \approx 0.60 \times 10^{-6} \text{ g}^2.$$

So the estimate $\hat{m}$ of m is seen as having been formed in a process with bias 0.082×10^{-3} g and standard deviation

$$\sqrt{\operatorname{var}_p \hat{m}} \approx \sqrt{\hat{\mathcal{E}}(E^2) - \hat{\mathcal{E}}(E)^2}$$
$$= 0.77 \times 10^{-3} \text{ g}.$$

Using the normal approximation, we then find that the interval [50.0022 g, 50.0052 g] is a realized approximate 95% average confidence interval for θ. This can be taken as a 95% uncertainty interval for θ.

If we had not incorporated the non-linear terms but used the normal-approximation method of Section 7.2 then we would have obtained the interval [50.0023 g, 50.0053 g]. Incorporating the non-linear terms has affected the location of the interval through the adjustment for bias, but has not noticeably affected the width of the interval. If we had not incorporated the non-linear terms but used the cumulants method instead, the interval obtained would have been [50.0025 g, 50.0051 g]. This interval is shorter because the parent distribution attributed to the most influential error, which is the error in $\hat{\epsilon}$, is uniform not normal.

Our subject in this chapter is the evaluation of measurement uncertainty in non-linear problems. This section has demonstrated that an approach based on higher-order analysis of the Taylor expansion is difficult, is of limited benefit and is quite possibly counter-productive. The next section describes a different approach to the construction of an interval and shows that, in simple situations, the Monte Carlo principle can be used to implement it.

8.3 Monte Carlo simulation of the measurement

A confidence interval for the target value in a non-linear problem can be constructed using the logic employed at the beginning of Section 6.2, where a candidate value for the target value $\theta \equiv \mathcal{F}(\theta_1, \ldots, \theta_m)$ was envisaged and then it was ascertained if the estimate x was consistent with this candidate value. In Section 6.2 we studied the error distribution, which did not depend on the values of the unknowns $\theta_1, \ldots, \theta_m$. Now the distribution of the error depends on these unknowns, and it becomes more convenient to work with the distribution of potential measurement results instead.

Consider the textbook case where the estimator of θ is $X \equiv \mathcal{F}(X_1, \ldots, X_m)$. Let the vector $\boldsymbol{\theta}_c \equiv (\theta_{1c}, \ldots, \theta_{mc})$ be a candidate for the true parameter vector $\boldsymbol{\theta} \equiv (\theta_1, \ldots, \theta_m)$, and let $q_{0.025}(X|\boldsymbol{\theta}_c)$ and $q_{0.975}(X|\boldsymbol{\theta}_c)$ denote the 0.025 and 0.975 quantiles of the distribution of X on the condition that $\boldsymbol{\theta}$ is equal to $\boldsymbol{\theta}_c$. If

$$q_{0.975}(X|\boldsymbol{\theta}_c) \leq x$$

then the value $\mathcal{F}(\boldsymbol{\theta}_c)$ is deemed too small to be a possible value for θ at a level of confidence of 97.5%. Similarly, if

$$q_{0.025}(X|\boldsymbol{\theta}_c) \geq x$$

then θ_c is deemed too large to be a possible value for θ at a level of confidence of 97.5%. As in Section 6.2, we predetermine to form a 95% confidence interval for θ as the intersection of a left-infinite 97.5% confidence interval and a right-infinite 97.5% confidence interval. This gives the following result.

Result 8.1 Suppose the target value is $\theta = \mathcal{F}(\boldsymbol{\theta})$, where $\boldsymbol{\theta} = (\theta_1, \ldots, \theta_m)$ and where each θ_i is unknown and has estimate x_i created during the measurement. Set $x = \mathcal{F}(\mathbf{x})$, where $\mathbf{x} = (x_1, \ldots, x_m)$. Let $\boldsymbol{\theta}_c$ denote a possible value of $\boldsymbol{\theta}$ and let $q_p(X|\boldsymbol{\theta}_c)$ denote the p-quantile of the parent distribution of x that would apply if $\boldsymbol{\theta}$ is equal to $\boldsymbol{\theta}_c$. The lower limit of a valid realized 95% confidence interval for θ is the smallest value of $\mathcal{F}(\boldsymbol{\theta}_c)$ for vectors $\boldsymbol{\theta}_c$ satisfying $q_{0.975}(X|\boldsymbol{\theta}_c) > x$. The upper limit is the largest value of $\mathcal{F}(\boldsymbol{\theta}_c)$ for vectors $\boldsymbol{\theta}_c$ satisfying $q_{0.025}(X|\boldsymbol{\theta}_c) < x$.

(Because of continuity, the limits can be taken as the smallest value of $\mathcal{F}(\boldsymbol{\theta}_c)$ for vectors $\boldsymbol{\theta}_c$ satisfying $q_{0.975}(X|\boldsymbol{\theta}_c) = x$ and the largest value of $\mathcal{F}(\boldsymbol{\theta}_c)$ for vectors $\boldsymbol{\theta}_c$ satisfying $q_{0.025}(X|\boldsymbol{\theta}_c) = x$.)

The interval is valid because jointly (i) the measurement had probability 0.95 of generating an estimate lying between $q_{0.025}(X|\boldsymbol{\theta})$ and $q_{0.975}(X|\boldsymbol{\theta})$ and (ii) applying the procedure to such an estimate will lead to an interval containing the target value $\mathcal{F}(\boldsymbol{\theta})$.

Example 8.1 (continued) Recall that the target value is $\theta = \theta_1^2 + \theta_2^2$ and the estimator is $X = X_1^2 + X_2^2$. The individual error variables $X_1 - \theta_1$ and $X_2 - \theta_2$ independently have the normal distribution with mean zero and variance σ^2. It follows that the variate X/σ^2 has the non-central chi-square distribution with two degrees of freedom and non-centrality parameter θ/σ^2 (Johnson et al., 1995).

Set $\boldsymbol{\theta}_c = (\theta_{1c}, \theta_{2c})$ and $\theta_c = \theta_{1c}^2 + \theta_{2c}^2$. So the figures $q_{0.025}(X|\boldsymbol{\theta}_c)$ and $q_{0.975}(X|\boldsymbol{\theta}_c)$ are equal to σ^2 multiplied by the 0.025 and 0.975 quantiles of the non-central chi-square distribution with two degrees of freedom and non-centrality parameter θ_c/σ^2. Figure 8.1 shows the region in the plane of θ_{1c}^2 and θ_{2c}^2 for which these two figures bound x with example values of $\sigma = 1$, $x_1 = 101.5$ and $x_2 = 99.3$. The region is a band between straight lines because the distribution of $X|\boldsymbol{\theta}_c$ depends on the candidate values θ_{1c} and θ_{2c} only through the sum of their squares, which is θ_c. The minimum and maximum values of $\mathcal{F}(\boldsymbol{\theta}_c) = \theta_c$ in this region are the values at the edges of the band. With these values of σ, x_1 and x_2, these extreme values are 137.93 and 245.54. So a realized 95% confidence interval for θ is [137.93, 245.54].

By construction, the success rate of this procedure is 0.95 whatever the values of θ_1, θ_2 and σ^2. Also, as we now show, the interval generated is slightly shorter than the interval (8.8) obtained using the linear approximation, which was the

8.3 Monte Carlo simulation of the measurement

Table 8.2. *Ratio of the width of the interval to the width of interval (8.8)*

x/σ^2	Ratio of widths
1	0.886
5	0.926
10	0.973
15.4	0.983
20	0.990
50	0.995
100	0.997
200	0.999

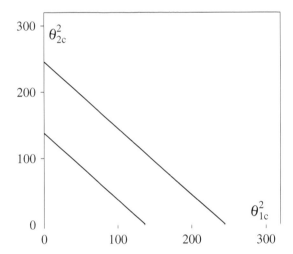

Figure 8.1 Region in parameter space consistent with the data and with the error model for $\sigma = 1$, $x_1 = 101.5$ and $x_2 = 99.3$.

preferred interval in Table 8.1. This approach therefore seems superior to any of those in Section 8.1 in this measurement situation.

The width of the interval generated is $(\lambda_{0.025} - \lambda_{0.975})\sigma^2$, where λ_p is the non-centrality parameter for which x/σ^2 is the p-quantile of the non-central chi-square distribution with two degrees of freedom. Interval (8.8) has width $7.84\sqrt{(x/\sigma^2)} \times \sigma^2$, so the ratio of the two widths is a function of x/σ^2 only. Table 8.2 shows this ratio for different values of x/σ^2. The italicized results are misleading, because they correspond to occasions where $\sqrt{(x/\sigma^2)} < 3.92$, i.e. $x/\sigma^2 < 15.37$, for which the lower limit of (8.8) would be negative and would be raised to zero. Such a value for x would only occur with a small target value θ.

In Example 8.1 the variate for a simulated measurement result, $X|\boldsymbol{\theta}_c$, had a distribution with a known form. This meant that the figures $q_{0.025}(X|\boldsymbol{\theta}_c)$ and

$q_{0.975}(X|\boldsymbol{\theta}_c)$ used in Result 8.1 could be found analytically, so that a valid realized confidence interval for θ could be calculated analytically. Usually such a solution will not be available and we will instead use a Monte Carlo method, as now described.

The Monte Carlo step

The Monte Carlo principle can be used to evaluate the figures $q_{0.025}(X|\boldsymbol{\theta}_c)$ and $q_{0.975}(X|\boldsymbol{\theta}_c)$. Initially we consider the textbook situation where each θ_i is a parameter of the measurement. The measurement is simulated under the condition that $\boldsymbol{\theta}_c$ is the unknown vector of parameter values. The simulated result, $\tilde{x}$, can be obtained in the following way.

1. Draw $\{\tilde{e}_1, \ldots, \tilde{e}_m\}$ from the joint distribution of $\{E_1, \ldots, E_m\}$ that would be applicable if $\boldsymbol{\theta}_c$ were the unknown parameter vector.
2. Calculate $\tilde{x} = \mathcal{F}(\theta_{1c} + \tilde{e}_1, \ldots, \theta_{mc} + \tilde{e}_m)$.

When the errors in each x_i are created independently, the first step simplifies to the following:

1. For $i = 1, \ldots, m$ draw $\tilde{e}_i$ from the distribution of E_i that would be applicable if $\boldsymbol{\theta}_c$ were the unknown parameter vector.

The distribution of potential measurement results is built up by applying this process many times, and the quantiles $q_{0.025}(X|\boldsymbol{\theta}_c)$ and $q_{0.975}(X|\boldsymbol{\theta}_c)$ are subsequently found by ordering the set of $\tilde{x}$ values. If $q_{0.025}(X|\boldsymbol{\theta}_c) < x < q_{0.975}(X|\boldsymbol{\theta}_c)$ then the value $\theta_c = \mathcal{F}(\boldsymbol{\theta}_c)$ must be included in the realized confidence interval for θ. This interval is created by applying this procedure over the m-dimensional space of $\boldsymbol{\theta}_c$ vectors. The limits of the final interval are the smallest and largest values included.

Example 8.4 Consider the measurement of $\theta = \theta_1 \theta_2$ when the error in the estimate of θ_1 is drawn from the uniform distribution with limits ± 4 and the error in the estimate of θ_2 is drawn from the uniform distribution with limits ± 3. Suppose the data are $x_1 = 15.1$ and $x_2 = 33.2$, so the measurement estimate is $x = x_1 x_2 = 501.32$. We seek to find bounds on $\theta_c = \theta_{1c} \theta_{2c}$ for values of θ_{1c} and θ_{2c} that are consistent with the datum x and the error models.

To simulate the measurement process at the candidate vector $(\theta_{1c}, \theta_{2c})$ we multiply a value drawn from $U(\theta_{1c} - 4, \theta_{1c} + 4)$ by a value drawn from $U(\theta_{2c} - 3, \theta_{2c} + 3)$. The distribution of potential measurement estimates is generated by repeating the process 10^6 times. Figure 8.2 shows the distributions obtained when $(\theta_{1c}, \theta_{2c}) = (14, 31)$ and $(\theta_{1c}, \theta_{2c}) = (17, 35)$, for example, and also indicates the 0.025 and 0.975 quantiles of these distributions. (The distributions differ by more

8.3 Monte Carlo simulation of the measurement

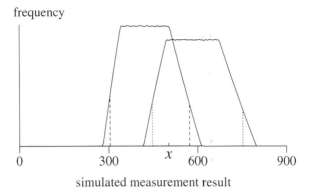

Figure 8.2 Distributions of measurement results that would be obtained if θ had been equal to (14, 31) (left) and (17, 35) (right). The discontinuous lines are situated at the 0.025 and 0.975 quantiles of the distributions.

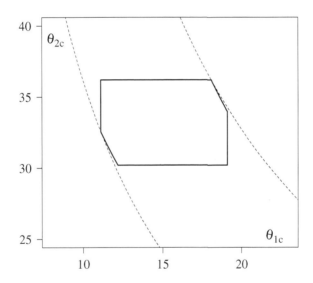

Figure 8.3 Closed curve: region in parameter space consistent with the data and the error model. Dashed lines: the functions $\theta_{1c}\theta_{2c} = 361$ and $\theta_{1c}\theta_{2c} = 655$.

than a simple translation because of the non-linearity of $\mathcal{F}$.) The central 95% of each of these distributions contains x, and the values of θ_{1c} and θ_{2c} are feasible given the data (x_1, x_2) and given the specified individual error distributions. So the values $\theta_c = 14 \times 31 = 434$ and $\theta_c = 17 \times 35 = 595$ are both to be included in the realized 95% confidence interval for θ.

Figure 8.3 shows the region in the $(\theta_{1c}, \theta_{2c})$ plane for which both the central 95% of the corresponding distribution contains x and the data-point (x_1, x_2) is permitted by the error models. The minimum and maximum values of θ_c in this region are 361 and 655. So a realized 95% confidence interval for θ is [361, 655]. Figure 8.3 also shows the functions $\theta_{1c}\theta_{2c} = 361$ and $\theta_{1c}\theta_{2c} = 655$.

The interval obtained may be compared with the interval [343, 660] obtained using the normal-approximation method and the narrower interval [358, 645] obtained the cumulants method.

The procedure may be summarized as follows. Monte Carlo simulation is used with a known candidate vector $(\theta_{1c}, \ldots, \theta_{mc})$ to obtain the distribution of potential measurement estimates that would exist if this vector was equal to the unknown true vector $(\theta_1, \ldots, \theta_m)$. The space of permissible candidate vectors is searched to find the extreme values of $\theta_c = \mathcal{F}(\theta_{1c}, \ldots, \theta_{mc})$ for which the central 95% of the corresponding distribution contains x. These extreme values of θ_c form the realization of a 95% confidence interval for θ. In this way, we are excluding all possible values of θ for which the result x is deemed too extreme to have occurred by chance.

Difficulties

The procedure just described suffers from at least two problems. The first is the practical problem that calculating the interval requires finding global extrema in a parameter space of dimension m. So although this approach is valid, it is somewhat impractical in the general case of sizeable m. The second problem relates to the fact that only the datum x is being assessed for compatibility with a candidate vector: the individual x_i data are not being considered. To see the effect of this, imagine that the parent distributions of the errors in Example 8.3 had unchanged variances but were normal, not uniform. These normal distributions are unbounded, so there are no longer obvious limits on the space of candidate vectors to be searched. We would find, for example, that the candidate vector $(\theta_{1c}, \theta_{2c}) = (5, 200)$ gave a distribution of potential results with central 95% covering x, so the figure 1000 would need to be included in the uncertainty interval. Yet it is clear than this candidate vector is highly inconsistent with the error models and the data vector $(x_1, x_2) = (15.1, 33.2)$. It follows that the interval obtained will be much wider than is appropriate: the procedure will be valid but will be very inefficient (in the statistical sense of the word). The next section describes how we can overcome this problem.

(This difficulty arises because of the non-linear relationship between the independent variates featuring in the *estimator equation* $X = X_1 X_2$. This difficulty does not exist in Example 8.1, which also involved normal errors, because the only relationship between independent variates there was additive: the estimator X was the sum of the independent variates X_1^2 and X_2^2.)

8.4 Monte Carlo simulation – an improved procedure

To overcome the second problem just described, where inappropriate values of θ_c are included in the interval, we must disqualify candidate vectors θ_c that are distant from the data vector $\mathbf{x}$ while keeping the success rate at least 0.95. This is achieved in the following way.

As in Section 8.3, let $q_p(X|\theta_c)$ denote the p-quantile of the distribution of X on the condition that the actual parameter vector is θ_c. Similarly, let $q_p(X_i|\theta_{ic})$ indicate the p-quantile of the distribution of X_i on the condition that θ_i is equal to θ_{ic}. Suppose that j of the x_i estimates are formed with unbounded error distributions and that the labelling is such that these are $x_1, \ldots, x_j$. We introduce a small probability δ, divide this probability into $j-1$ parts $\delta_1, \ldots, \delta_{j-1}$ satisfying

$$\sum_{i=1}^{j-1} \delta_i = \delta, \tag{8.25}$$

and then restrict ourselves to searching the space of candidate vectors θ_c satisfying

$$q_{\delta_i/2}(X_i|\theta_{ic}) < x_i < q_{1-\delta_i/2}(X_i|\theta_{ic}), \qquad i = 1, \ldots, j-1. \tag{8.26}$$

In this way, the j unbounded errors are prevented from combining to generate a result that is consistent with x but inconsistent with the x_i data. (Applying the condition to $j-1$ of the unbounded errors controls the remaining unbounded error also.) The interval is then formed by including the θ_c values for all vectors in this space satisfying

$$q_{0.025-\delta/2}(X|\theta_c) < x < q_{0.975+\delta/2}(X|\theta_c).$$

Again, the limits of the interval quoted are the extreme values included. If we are to maintain our focus on the efficient estimation of θ then δ must be considerably less that 0.05. A suitable choice is $\delta = 0.001$.

Let us now demonstrate the validity of this procedure. For each $i = 1, \ldots, j-1$, the measurement of θ_i had probability $1-\delta_i$ of generating a value x_i lying between $q_{\delta_i/2}(X_i|\theta_i)$ and $q_{1-\delta_i/2}(X_i|\theta_i)$. Also, the overall measurement had probability $0.95+\delta$ of generating a value x lying between $q_{0.025-\delta/2}(X|\theta)$ and $q_{0.975+\delta/2}(X|\theta)$. No matter what dependence existed between the occurrence of these j events, the probability that not all of them occurred was no greater than

$$\sum_{i=1}^{j-1} \delta_i + \{1 - (0.95 + \delta)\} = 0.05$$

(by the Bonferroni inequality described in Chapter 10), which means that the probability that they all occurred was at least 0.95. Thus the measurement had probability at least 0.95 of generating results that would lead to the calculation of an interval containing the target value θ. Therefore, the procedure is valid.

This analysis permits the statement of the following result.

Result 8.2 Suppose the target value is $\theta = \mathcal{F}(\boldsymbol{\theta})$, where $\boldsymbol{\theta} = (\theta_1, \ldots, \theta_m)$ and where each θ_i is unknown and has estimate x_i created during the measurement. Set $x = \mathcal{F}(\mathbf{x})$, where $\mathbf{x} = (x_1, \ldots, x_m)$. Let $\boldsymbol{\theta}_c = (\theta_{1c}, \ldots, \theta_{mc})$ denote a general possible value of $\boldsymbol{\theta}$, let $q_p(X|\boldsymbol{\theta}_c)$ denote the p-quantile of the parent distribution of x on the condition that the true parameter vector is equal to $\boldsymbol{\theta}_c$ and let $q_p(X_i|\theta_{ic})$ denote the p-quantile of the distribution of X_i on the condition that θ_i is equal to θ_{ic}. Suppose that the estimates $x_{j+1}, \ldots, x_m$ are created in processes with bounded errors. Let δ and $\delta_1, \ldots, \delta_{j-1}$ be predetermined tiny probabilities such that $\delta = \sum_{i=1}^{j-1} \delta_i$. Then the lower limit of a valid realized 95% confidence interval for θ is the smallest value of $\mathcal{F}(\boldsymbol{\theta}_c)$ for vectors $\boldsymbol{\theta}_c$ satisfying both $q_{0.975+\delta/2}(X|\boldsymbol{\theta}_c) > x$ and (8.26). The upper limit is the largest value of $\mathcal{F}(\boldsymbol{\theta}_c)$ for vectors $\boldsymbol{\theta}_c$ satisfying both $q_{0.025-\delta/2}(X|\boldsymbol{\theta}_c) < x$ and (8.26).

The use of this result is illustrated in the following example.

Example 8.5 The density ρ of an object is estimated by measuring the mass m and volume V of the object. The target value is $\rho = m/V$. Fluctuations in temperature and pressure can be ignored, so m and V are seen as parameters of the measurement.

Suppose the mass and volume are independently measured using processes with negligible biases but with Gaussian random errors having standard deviations 0.04 g and 0.07 cm^3. Then the statistical component of the measurement model is

$$\hat{m} \leftarrow N(m, 0.04^2) \text{ g},$$
$$\hat{V} \leftarrow N(V, 0.07^2) \text{ cm}^3.$$

Let the estimates obtained be $\hat{m} = 3.81$ g and $\hat{V} = 4.17$ cm^3. The measurement result is $\hat{\rho} = \hat{m}/\hat{V} = 0.914$ g cm^{-3}.

In this example $j = 2$, so $\delta_1 = \delta$. We choose to apply Result 8.2 with $\delta = 0.001$, and we identify m with θ_1. The $0.001/2$ and $1 - 0.001/2$ quantiles of the standard normal distribution are ± 3.29, so the part of the plane of candidate mass-volume vectors that we search is the band defined by

$$\hat{m} - 0.132 < m_c < \hat{m} + 0.132, \tag{8.27}$$

with the units being grams. This is depicted in Figure 8.4. The simulation of a measurement result for a candidate pair (m_c, V_c) is carried out by dividing a value drawn from the distribution $N(m_c, 0.04^2)$ by a value independently drawn from the distribution $N(V_c, 0.07^2)$. This process is carried out 10^6 times, and the set of values obtained is sorted from lowest to highest. Because $\delta = 0.001$ we wish to find the 0.0245 and 0.9755 quantiles of the potential distribution of measurement results with

8.4 Monte Carlo simulation – an improved procedure

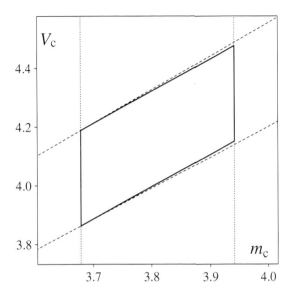

Figure 8.4 Closed curve: region in parameter space consistent with the data and the error model. Dashed lines: the functions $m_c/V_c = 0.878$ g cm^{-3} and $m_c/V_c = 0.952$ g cm^{-3}. Dotted lines: the functions $m_c = 3.678$ g cm^{-3} and $m_c = 3.942$ g cm^{-3}.

this candidate vector. These are closely approximated by the elements in positions 24 500 and 975 500 of this ordered set, which cover $100(0.95+\delta)\% = 95.1\%$ of this set. If $\hat{\rho}$ lies between these values then the realized interval must include the value $\rho_c = m_c/V_c$.

Consider the plane of (m_c, V_c) points. Figure 8.4 shows the region within the band defined by (8.27) and for which the central 95.1% of the corresponding distribution contains $\hat{\rho}$. The minimum and maximum values of ρ_c in this region are 0.878 g cm^{-3} and 0.952 g cm^{-3}, so these are the limits of a realized 95% confidence interval for ρ. Figure 8.4 also shows the functions $m_c/V_c = 0.878$ g cm^{-3} and $m_c/V_c = 0.952$ g cm^{-3}.

The interval obtained may be compared with the interval [0.878, 0.949] g cm^{-3} obtained using either the normal-approximation method or the cumulants method.

(If we revert to the method of Section 8.3 by setting $\delta = 0$ then we find that the calculated interval must contain the value 0.5 g cm^{-3}, for example, which arises with the candidate pair $(m_c, V_c) = (0.1, 0.2)$.)

The method can take into account the relative influences of the errors in the $j-1$ estimates through the unequal division of δ, which is possible when $j \geq 3$. The length of the interval will be reduced by giving larger values of δ_i to the more influential errors. However, we are not at liberty to choose the δ_i values after observing the data so as to quote the shortest interval. That practice would be analogous to the practice of 'cherry picking' described in Chapter 10. One possibility would be to predetermine to take the figure

$$a_i = \left| \mathcal{F}(x_1, \ldots, x_i + \sigma_p(e_i), \ldots, x_m) - \mathcal{F}(x_1, \ldots, x_m) \right| \quad (8.28)$$

as a measure of the importance of the ith estimate and then to define $\delta_k = a_k \delta / \sum_{i=1}^{j-1} a_i$.

An illusory short-cut

The Monte Carlo methods described so far in this chapter suffer from the problem that computation of the interval becomes difficult as m increases. A tempting simplification is only to evaluate the distribution of measurement errors at $\boldsymbol{\theta}_c = \mathbf{x}$ and then to treat this distribution as being applicable at all values of $\boldsymbol{\theta}_c$ in the neighbourhood of $\boldsymbol{\theta}$. If $e_{0.025}(\mathbf{x})$ and $e_{0.975}(\mathbf{x})$ are the 0.025 and 0.975 quantiles of this distribution then the interval quoted will be $[x - e_{0.975}(\mathbf{x}), x - e_{0.025}(\mathbf{x})]$, which is the interval

$$[2x - q_{0.975}(X|\mathbf{x}), \; 2x - q_{0.025}(X|\mathbf{x})].$$

This short-cut method is based on the assumption that the error distribution does not change significantly with $\boldsymbol{\theta}_c$ in the vicinity of $\boldsymbol{\theta}$ – and this assumption is a large part of the linear approximation in disguise! Consequently, we might expect this method to perform poorly when the linear approximation is inadequate.

Example 8.1 (continued) The target value is $\theta = \theta_1^2 + \theta_2^2$ and the estimate is $x = x_1^2 + x_2^2$. The estimator X is such that X/σ^2 has the non-central chi-square distribution with two degrees of freedom and non-centrality parameter θ/σ^2. So the interval calculated by the short-cut method just described is

$$[2x - b(x)\sigma^2, \; 2x - a(x)\sigma^2],$$

where $a(x)$ and $b(x)$ are the 0.025 and 0.975 quantiles of the non-central chi-squared distribution with two degrees of freedom and non-centrality parameter x/σ^2. As earlier, simulations were carried out with $\theta_1 = \theta_2 = 10$ and with different values of σ. (For numerical reasons, the simulations were not as extensive as those giving the results in Table 8.1. Here there were only 10^5 replications at each value of σ, and the lower values of σ were not considered.) Table 8.3 shows the success rate with this method. The method performs considerably worse than each of the methods examined in Table 8.1.

Table 8.3. *Success rate with the short-cut simulation method*

σ	Success rate
0.5	0.948
1	0.942
2	0.923
3	0.895

8.5 Monte Carlo simulation – variant types of interval

Let us now consider how the concepts of average confidence and conditional confidence discussed in Chapter 5 affect the procedure described in Section 8.4.

Average confidence interval

In the simulations discussed so far in this chapter, each θ_i is an unknown parameter of the measurement and each x_i is generated in the measurement. However, in some measurements one or more of the x_i estimates has a fixed value (as in Example 1.1, where $\hat{\delta} = 0$, and Example 8.3, where $\hat{m}_{\text{ref}} = 50$ g and $\hat{\epsilon} = 0$). If x_i is fixed and the error e_i is to be treated probabilistically then θ_i must be seen as having a parent distribution, as in the idea of 'average confidence' described in Section 5.1. Instead of repeatedly generating a potential estimate $\tilde{x}_i$ we now must repeatedly generate a $\tilde{\theta}_i$ value that is consistent with x_i. The required success rate of 95% must therefore refer to a success rate in measurements of this type with different values of θ_i. The interval we obtain is a realized average confidence interval.

We suppose that $x_1, \ldots, x_k$ are estimates of quantities $\theta_1, \ldots, \theta_k$ which are parameters of the measurement and that the remaining estimates $x_{k+1}, \ldots, x_m$ are fixed. As in Section 8.4, we suppose that the estimates formed using unbounded errors are $x_1, \ldots, x_j$ (with $j \leq k$). Again we involve a probability δ linked to the extremeness allowed for the measurement estimate x, and again we involve smaller probabilities linked to the extremeness allowed for $x_1, \ldots, x_{j-1}$. Now, however, we associate another proportion δ' with the joint extremeness allowed for $\theta_{k+1}, \ldots, \theta_m$, and we require that

$$\delta' + \sum_{i=1}^{j-1} \delta_i = \delta, \qquad (8.29)$$

which is a generalization of (8.25).

The unknown vector of parameters is now $\boldsymbol{\theta} \equiv (\theta_1, \ldots, \theta_k)$ and a candidate vector for $\boldsymbol{\theta}$ is now $\boldsymbol{\theta}_c = (\theta_{1c}, \ldots, \theta_{kc})$. These are vectors of length k, not m. We again restrict our attention to the space of candidate vectors for which (8.26) holds. For such a vector we:

1. draw $\{\tilde{e}_1, \ldots, \tilde{e}_k\}$ from the joint distribution of $\{E_1, \ldots, E_k\}$ that would be applicable if $\boldsymbol{\theta}_c$ was the unknown vector of k parameters,
2. calculate $\tilde{x} = \mathcal{F}(\theta_{1c} + \tilde{e}_1, \ldots, \theta_{kc} + \tilde{e}_k, x_{k+1} \ldots, x_m)$,

and build up the distribution of potential measurement results for this candidate vector by applying this process many times. If x lies between the $0.025 - \delta/2$ and $0.975 + \delta/2$ quantiles of this distribution then the vector $\boldsymbol{\theta}_c$ is accepted as being consistent with the data.

For each accepted candidate vector $\boldsymbol{\theta}_c$, we examine the distribution of the target values that are consistent with the fixed estimates $x_{k+1}, \ldots, x_m$ and with the parent distributions of the errors $e_{k+1}, \ldots, e_m$. (This can be regarded as the conditional probability distribution of a variate Θ_c in all measurement problems of this type.) This distribution is found by repeating the following steps.

1. Draw $\{\tilde{e}_{k+1}, \ldots, \tilde{e}_m\}$ from the joint distribution of $\{E_{k+1}, \ldots, E_m\}$ that would be applicable if $\boldsymbol{\theta}_c$ were the unknown vector of k parameters.
2. Calculate $\tilde{\theta}_c = \mathcal{F}(\theta_{1c}, \ldots, \theta_{kc}, x_{k+1} - \tilde{e}_{k+1}, \ldots, x_m - \tilde{e}_m)$.

On average, the target values in $(1 - \delta') \times 100\%$ of these measurements are to be included in the average confidence interval for θ. This is ensured by including the $\delta'/2$ and $1 - \delta'/2$ quantiles of the simulated dataset. The interval is built up by searching the space of candidate vectors and applying this procedure with every vector $\boldsymbol{\theta}_c$ that is accepted. The limits of the final interval are the lowest and highest values that are included.

Let us see how the interval obtained is a valid realized average 95% confidence interval. The measurement had probability at least $1 - \sum_{i=1}^{j-1} \delta_i$ of generating estimates $x_i, \ldots, x_{j-1}$ satisfying

$$q_{\delta_i/2}(X_i|\theta_i) < x_i < q_{1-\delta_i/2}(X_i|\theta_i) \quad i = 1, \ldots, j-1$$

and had probability $0.95 + \delta$ of generating a result x lying in the interval

$$[\theta^* + q_{0.025-\delta/2}(X|\boldsymbol{\theta}), \theta^* + q_{0.975+\delta/2}(X|\boldsymbol{\theta})],$$

where $\theta^* \equiv \mathcal{F}(\theta_1, \ldots, \theta_k, x_{k+1} \ldots, x_m)$. No matter what dependence existed between the occurrence of these j events, the probability that not all of them occurred was no greater than

$$\sum_{i=1}^{j-1} \delta_i + \{1 - (0.95 + \delta)\} = 0.05 - \delta',$$

8.5 Monte Carlo simulation – variant types of interval   147

which means that the probability that they all occurred was at least $0.95 + \delta'$. So in at least $(0.95 + \delta') \times 100\%$ of measurements of this type we will obtain data with which the target value and the actual parameter vector $\boldsymbol{\theta}$ would be deemed to be consistent. Also, in $(1 - \delta') \times 100\%$ of measurements of this type the target value $\theta = \mathcal{F}(\theta_1, \ldots, \theta_m)$ will be among the values included in the interval. Therefore at least 95% of measurements of this type will result in data from which an interval containing the target value will subsequently be calculated. By Definition 5.1, this interval is a realized 95% average confidence interval for θ.

This gives the following result.

Result 8.3 The target value is $\theta = \mathcal{F}(\theta_1, \ldots, \theta_m)$ where each θ_i is unknown and has estimate x_i. Suppose that

1. the estimates $x_1, \ldots, x_j$ are created from the values $\theta_1, \ldots, \theta_j$ in processes with unbounded errors,
2. the estimates $x_{j+1}, \ldots, x_k$ are created from the values $\theta_{j+1}, \ldots, \theta_k$ in processes with bounded errors, and
3. the estimates $x_{k+1}, \ldots, x_m$ have fixed values.

Set $x = \mathcal{F}(x_1, \ldots, x_m)$. Let δ, δ' and $\delta_1, \ldots, \delta_{j-1}$ be predetermined small probabilities such that $\delta = \delta' + \sum_{i=1}^{j-1} \delta_i$, and let $q_{\delta'/2}(\Theta_c)$ and $q_{1-\delta'/2}(\Theta_c)$ be the $\delta'/2$ and $1 - \delta'/2$ quantiles of the distribution of the variate

$$\Theta_c = \mathcal{F}(\theta_{1c}, \ldots, \theta_{kc}, x_{k+1} - E_{k+1}, \ldots, x_m - E_m),$$

where θ_{ic} is a possible value for θ_i and E_i is a variate whose distribution is the parent distribution of the error $x_i - \theta_i$. Define $\boldsymbol{\theta}_c \equiv (\theta_{1c}, \ldots, \theta_{kc})$, let $q_p(X|\boldsymbol{\theta}_c)$ denote the p-quantile of the parent distribution of x on the condition that the true abbreviated vector $(\theta_1, \ldots, \theta_k)$ is equal to $\boldsymbol{\theta}_c$, and let $q_p(X_i|\theta_{ic})$ denote the p-quantile of the distribution of X_i on the condition that θ_i is equal to θ_{ic}. The lower limit of a valid realized 95% average confidence interval for θ is the smallest value of $q_{\delta'/2}(\Theta_c)$ for vectors $\boldsymbol{\theta}_c$ satisfying both $q_{0.975+\delta/2}(X|\boldsymbol{\theta}_c) > x$ and (8.26). The upper limit is the largest value of $q_{1-\delta'/2}(\Theta_c)$ for vectors $\boldsymbol{\theta}_c$ satisfying both $q_{0.025-\delta/2}(X|\boldsymbol{\theta}_c) < x$ and (8.26).

Result 8.3 reduces to Result 8.2 when $k = m$, where δ' will be set to 0.

Choosing δ is now more difficult than in Section 8.4. The choice should be made with regard to the relative influences of the set of errors in the fixed estimates and the set of errors in the unfixed estimates. If the errors in $x_1, \ldots, x_k$ are dominant then δ should be set close to the small value that was applicable in Section 8.4, but if the errors in $x_{k+1}, \ldots, x_m$ are dominant then δ should be set close to its maximum value of 0.05. Therefore, we might calculate

$$\delta \approx \frac{\sum_{i=k+1}^{m} a_i}{\sum_{i=1}^{m} a_i} \times 0.05, \tag{8.30}$$

with a_i as defined in (8.28). In this way, when the overall parent error variance is associated evenly between the two types of estimates the accepted 'Type I error probability' of 0.05 is partitioned evenly.

Consider again Example 8.3.[1]

Example 8.3 (revisited) A weight of density ρ and unknown mass m is calibrated against a reference weight of density ρ_{ref} and mass m_{ref} using a balance that is operated in air of density ρ_{air}. Both masses are nominally equal to 50 g. The measurand equation is

$$\theta = (\theta_1 + \theta_4) \frac{1 - \theta_2/\theta_3}{1 - \theta_2/(\theta_3 + \theta_5)},$$

with $\theta = m$ and $(\theta_1, \theta_2, \theta_3, \theta_4, \theta_5) = (s, \rho_{\text{air}}, \rho_{\text{ref}}, m_{\text{ref}}, \epsilon)$. The estimates are

$$\begin{aligned}
x_1 &= \hat{s} &&= 3.8 \times 10^{-3} \text{ g}, \\
x_2 &= \hat{\rho}_{\text{air}} &&= 1.2 \times 10^3 \text{ g m}^{-3}, \\
x_3 &= \hat{\rho}_{\text{ref}} &&= 8.2 \times 10^6 \text{ g m}^{-3}, \\
x_4 &= \hat{m}_{\text{ref}} &&= 50 \text{ g}, \\
x_5 &= \hat{\epsilon} &&= 0 \text{ g m}^{-3}
\end{aligned}$$

and the error model is

$$E_1 \sim \text{N}\left(0, \left(0.1 \times 10^{-3}\right)^2\right),$$

$$E_2 \sim \text{U}(-0.3 \times 10^3, \ 0.3 \times 10^3),$$

$$E_3 \sim \text{U}(-0.5 \times 10^6, \ 0.5 \times 10^6),$$

$$E_4 \sim \text{N}\left(0, \left(5 \times 10^{-6}\right)^2\right),$$

$$E_5 \sim \text{U}(-1.5 \times 10^6, \ 1.5 \times 10^6).$$

Equation (8.28) gives $(a_1, a_2, a_3, a_4, a_5) = (10^{-4}, 0, 0, 5 \times 10^{-6}, 7.0 \times 10^{-4})$. The two estimates $\hat{m}_{\text{ref}}$ and $\hat{\epsilon}$ are fixed, so $k = 5 - 2 = 3$. Therefore, from (8.30), we choose $\delta = 0.044$.

The candidate vector $(s_c, \rho_{\text{air},c}, \rho_{\text{ref},c})$ is of length $k = 3$, and the candidate version of the measurand equation is

---

[1] As has been noted, this example is based on one found in Supplement 1 to the *Guide* (2008b, Example 9.3). That supplement takes the view that probability distributions can be directly attributed to each θ_i quantity. Consequently, the formulation of the Monte Carlo approach to uncertainty evaluation put forward there differs considerably from the formulation in this book. The approach put forward in that supplement can lead to unacceptable results (e.g. Hall (2008), Willink (2010a)).

8.5 Monte Carlo simulation – variant types of interval

$$m = (s_c + m_{\text{ref}}) \frac{1 - \rho_{\text{air,c}}/\rho_{\text{ref,c}}}{1 - \rho_{\text{air,c}}/(\rho_{\text{ref,c}} + \epsilon)}.$$

The estimate of ϵ is zero, so the corresponding potential distribution of measurement results is unaffected by the values of $\rho_{\text{air,c}}$ and $\rho_{\text{ref,c}}$. So, in effect, the search of the space of candidate vectors is carried out in only one dimension, which is the 's' dimension. This makes it easy to find the set of candidate vectors that satisfy the conditions $q_{0.025-\delta/2}(X|\theta_c) < x$ and $x < q_{0.975+\delta/2}(X|\theta_c)$ implied in Result 8.3. The $0.975 + \delta/2$ and $0.025 - \delta/2$ quantiles of the standard normal distribution are ± 2.75, so the candidate vectors $(s_c, \rho_{\text{air,c}}, \rho_{\text{ref,c}})$ that satisfy these conditions are those for which

$$s_c - 2.75 \times 0.0001 \leq 0.0038 \leq s_c + 2.75 \times 0.0001,$$

which are those for which $0.003525 \leq s_c \leq 0.004075$. Only one of the $k = 3$ errors in the estimates created in the measurement is unbounded, so $j = 1$. Therefore, condition (8.26) is empty and is satisfied, which means that all of these candidate vectors are acceptable.

The fact that $j = 1$ means that the sum in (8.29) is empty. Thus $\delta' = \delta$, so $\delta' = 0.044$. The variate for m corresponding to a candidate value $(s_c, \rho_{\text{air,c}}, \rho_{\text{ref,c}})$ is

$$M_c = (s_c + \hat{m}_{\text{ref}} - E_4) \frac{1 - \rho_{\text{air,c}}/\rho_{\text{ref,c}}}{1 - \rho_{\text{air,c}}/(\rho_{\text{ref,c}} - E_5)}.$$

(This corresponds to the variate Θ_c in Result 8.3.) The smallest value of $q_{\delta'/2}(M_c)$ is obtained when $\rho_{\text{air,c}}$ is maximized, $\rho_{\text{ref,c}}$ is minimized and s_c is minimized. The largest value of $q_{1-\delta'/2}(M_c)$ is obtained without changing the values of $\rho_{\text{air,c}}$ and $\rho_{\text{ref,c}}$ but with s_c maximized. Monte Carlo analysis with 10^6 trials shows that these extreme values of $q_{\delta'/2}(M_c)$ and $q_{1-\delta'/2}(M_c)$ are approximately 50.0020 and 50.0063. So a realized valid evaluated 95% average confidence interval for m is [50.0020 g, 50.0063 g]. This interval is a 95% interval of measurement uncertainty for m.

The interval calculated is substantially longer that the interval [50.0022 g, 50.0052 g] obtained using the non-linear terms in Section 8.2. This is partly explained by the fact that the procedure is constructed for validity, and so it will err on the side of conservatism. In particular, we see that the limits of the interval here have eventually been formed using the maximum possible value for ρ_{air} and the minimum possible value for ρ_{ref}, without involving the minimum value for ρ_{air} and the maximum value for ρ_{ref}. The conservatism is a by-product of the fact that we require the interval to obey a strict interpretation, this interpretation being related to the success rate of the procedure used.

As explained, it is not strictly permissible to try several values of δ and then choose the one giving the shortest interval; δ must be chosen without regard to the data if the overall procedure can be seen to have probability at least 0.95 of being successful. In this situation, changing the value of δ would make very little difference to the result.

This section has presented the most complicated procedure to be found in this book. The development of this procedure represents a logical response to the fact that there are different types of θ_i quantities featuring in the general measurand equation $\theta = \mathcal{F}(\theta_1, \ldots, \theta_m)$. Some of these quantities are parameters with estimates created within the measurement, while some may be parameters with fixed estimates. And – as shall be restated in the next paragraph – some are to be seen as quantities that themselves have values created within the measurement. The procedure just described displays some of this complexity, even though it happens to involve a search of parameter space in only one dimension. The reader might be pleased to learn that the analysis beyond this point in the book is more straightforward.

Conditional confidence interval

In many measurement problems, some of the θ_i values will be created during the measurement process. For example one might be the value of an environmental quantity like temperature. In such a case, simulation of a measurement should strictly include the creation of this kind of θ_i value. We would need to know the distribution of the corresponding variate Θ_i in order to simulate the measurement. As a consequence, the proper simulation of the full measurement would become difficult. However, our interest is in calculating a confidence interval conditional on the events that we know have occurred, and one of these is the event $\Theta_i = \theta_i$, even though θ_i is unknown. So the measurement need only be simulated from the point where a candidate value for θ_i is fixed. The interval then obtained is a realized conditional confidence interval as described in Section 5.2.

Many measurement problems will fall into this category. Results 8.1–8.3 are applicable in these problems but the adjective 'conditional' should be added to the description of the confidence interval obtained.

8.6 Summarizing comments

This chapter has presented two different approaches to the evaluation of measurement uncertainty when non-linearity in the measurand equation $\theta = \mathcal{F}(\theta_1, \ldots, \theta_m)$ is taken into account. The reader will have observed greater computational and conceptual complexity here than in Chapter 7, and might reasonably conclude that the methods described in this chapter should be used as the exception rather than the rule. A further difficulty is the requirement for uncertainty information to be transferable from one measurement to another. When we depart from the pattern exhibited in Chapter 7 of quoting intervals that are symmetric about the

8.6 Summarizing comments

measurement result x, in effect we introduce an offset that must be acknowledged in the storage of the uncertainty information.

The Monte-Carlo-based methodology presented in this chapter is more complicated than that described in Section 7.4, where the distributions involved were fully known and the idea of searching a space of candidate parameters did not apply. The basic idea behind the methodology here is that of simulating the measurement process under different possible sets of conditions so that, whatever the true conditions may be, the whole measurement process has probability 0.95 of being successful. That is the principle of design on which this book is based. The uncertainty interval that results then has a meaningful interpretation. The methodology is unwieldy, and a simplified method of analysis might seem to be of advantage. But what would be the justification for a simpler method if the interval obtained both (i) lacked a proper interpretation in itself and (ii) was not necessarily numerically similar to an interval that did possess a proper interpretation? With this question in mind, we do not present any simpler Monte Carlo method for the analysis of non-linear problems.

9
Uncertainty information fit for purpose

The idea of seeing systematic errors as originating from parent distributions is a key feature in the methods of uncertainty evaluation given in the previous two chapters. An associated idea is to report a single total error variance, where the components due to random and systematic errors are combined. Although these ideas have merit, some dangers exist. One is that they can lead to a loss of distinction between the effects of the two types of error: random errors affect measurement results individually and systematic errors affect them collectively. Potentially important information is lost if all that is recorded are details of a single overall error distribution. This chapter discusses this difficulty and related issues, and describes some realistic and important situations where this loss of information might occur.

Section 9.1 demonstrates how the information contained in a single variance can be insufficient and advocates the additional use of a table of error components instead of a covariance matrix. Section 9.2 gives an alternative formulation of the law of propagation of error that makes use of such a table. Section 9.3 illustrates these ideas in the estimation of quantities that are derived from measurable functions. Section 9.4 discusses measurement in the routine testing of products, which is an area of application where bounds on the error seem more appropriate than a variance. Section 9.5 discusses the underlying concept that errors endure for different lengths of time.

9.1 Information for error propagation

One basic principle of measurement is the idea that information about uncertainty should be stored in a way that facilitates the evaluation of uncertainty in subsequent measurements. If the only information stored is an overall parent variance then this principle is violated. This is easily illustrated by considering the use of a single device when measuring two masses, θ_1 and θ_2. Let the measurements follow the equations

9.1 Information for error propagation

$$x_1 = \theta_1 + e_1, \qquad e_1 = e_{\text{ran},1} + e_{\text{fix}},$$
$$x_2 = \theta_2 + e_2, \qquad e_2 = e_{\text{ran},2} + e_{\text{fix}},$$

where $e_{\text{ran},1}$ and $e_{\text{ran},2}$ are random errors and e_{fix} is a fixed error associated with the use of the device. Suppose that the parent variance for both the random errors is σ_{ran}^2 and that the parent variance for the fixed error is σ_{fix}^2. Consider the function $\theta = \theta_1 + \theta_2$, as if the mass of some object is to be measured by adding the masses of its components. The measurement result will be

$$\begin{aligned} x &= x_1 + x_2 \\ &= \theta + e_{\text{ran},1} + e_{\text{ran},2} + 2e_{\text{fix}}, \end{aligned}$$

and the parent variance that should be attributed to the overall measurement error $e = x - \theta$ is

$$\sigma_{\text{correct}}^2 = 2\sigma_{\text{ran}}^2 + 4\sigma_{\text{fix}}^2.$$

This is the figure that we would like to calculate. However, if the only uncertainty information recorded with x_i is the parent variance of e_i, which is $\sigma_i^2 = \sigma_{\text{ran}}^2 + \sigma_{\text{fix}}^2$, then the figure $\sigma_{\text{correct}}^2$ cannot be obtained. Simply forming the sum

$$\sigma_1^2 + \sigma_2^2 = 2\sigma_{\text{ran}}^2 + 2\sigma_{\text{fix}}^2$$

gives a figure that is smaller, and so the corresponding uncertainty interval should be seen as invalid. We may conclude that if measurement uncertainty is expressed as a single variance then measurement estimates of different quantities obtained using the same device may not be added!

A different issue is observed with the function $\theta = \theta_1 - \theta_2$, which arises, for example, when a tare mass is measured by subtracting the mass of the packaging from the gross mass. The measurement result will be

$$\begin{aligned} x &= x_1 - x_2 \\ &= \theta + e_{\text{ran},1} - e_{\text{ran},2}, \end{aligned}$$

so the variance we wish to calculate is now $\sigma_{\text{correct}}^2 = 2\sigma_{\text{ran}}^2$. Again, if the only uncertainty information recorded alongside x_i is σ_i^2 we cannot calculate the correct result. In this case, forming the sum $\sigma_1^2 + \sigma_2^2$ gives a figure that is too large by $2\sigma_{\text{fix}}^2$. Although valid, the corresponding uncertainty interval will be unnecessarily long.

It is clear that these problems arise because the errors in x_1 and x_2 have not been generated independently. The correct variances can be calculated if sufficient information about the dependence is also stored, but what form should this information take?

A covariance matrix or a table of contributing variances?

In both the addition and subtraction problems above, the summation of σ_1^2 and σ_2^2 is an application of the law of propagation of error (7.11) for a linear function $\theta = \mathcal{F}(\theta_1, \ldots, \theta_m)$ that has uncorrelated error variates. The reason that incorrect results are obtained is that the error variates are correlated. Incorporating the appropriate parent covariance $\text{cov}_p(e_1, e_2) = \sigma_{\text{fix}}^2$ will give the intended results, the quantity $\partial \mathcal{F}/\partial x_1 \times \partial \mathcal{F}/\partial x_2$ in (7.11) being $+1$ and -1 in the two cases. We would also find the same to be true for general $\mathcal{F}$ and general m.

So fixed and random errors propagate in different ways when the same technique or device is used to measure two or more contributing θ_i quantities, but this difficulty is overcome by incorporating the off-diagonal entries in the covariance matrix implied in (7.11).

Example 6.1 (continued, using (7.23) not (7.27)) The distance d between the two points in the plane is measured. The estimate of the distance between the two points is expressible as $\hat{d}(\hat{r}_1, \hat{r}_2, \hat{\phi})$. The estimate is $\hat{d} = 39.14$ m, and the partial derivatives are $\partial \hat{d}/\partial \hat{r}_1 = -0.368$, $\partial \hat{d}/\partial \hat{r}_2 = 0.759$, $\partial \hat{d}/\partial \hat{\phi} = 50.95$ m rad^{-1}. Table 7.1 shows that the parent variances of the three errors are

$$\text{var}_p(\hat{r}_1 - r_1) \approx 0.022 \text{ m}^2,$$
$$\text{var}_p(\hat{r}_2 - r_2) = 0.034 \text{ m}^2,$$
$$\text{var}_p(\hat{\phi} - \phi) = 2.50 \times 10^{-5} \text{ rad}^2$$

and that the only non-zero parent covariance is $\text{cov}_p(\hat{r}_1 - r_1, \hat{r}_2 - r_2) = 0.017$ m^2. Applying the law of propagation of error without the covariance term gives an incorrect result [38.56, 39.72] m but including the covariance term gives a correct interval [38.59, 39.69] m, which is (7.9).

So the provision of a covariance matrix avoids the kind of problem described above. However, establishing this matrix requires knowledge of the components of error that affect groups of θ_i values collectively. Therefore, if no record is kept of the individual elements of error contributing to a parent variance $\sigma_p^2(e_i)$ then no parent covariance involving e_i can be calculated. For example if the radial distances r_1 and r_2 in Example 6.1 had earlier been measurands in their own right and if the only uncertainty information recorded about them had been $\text{var}_p(\hat{r}_1 - r_1)$ and $\text{var}_p(\hat{r}_2 - r_2)$ then the correct result [38.59, 39.69] m could not have been obtained. We would not have been able to calculate the parent covariance $\text{cov}_p(\hat{r}_1 - r_1, \hat{r}_2 - r_2) = 0.017$ m^2.

A second difficulty with the idea of storing the uncertainty information in a covariance matrix is the potential size of this matrix. An $m \times m$ covariance matrix has potentially $m(m+1)/2$ distinct entries, and m might be large, as we shall see in Section 9.2. Often it will be easier to store the uncertainty information in a table with one row for each of the m estimates and with various columns for different sources of error. The column headings will reflect the source of the error, so such a table maintains explanatory information about the measurement process. The covariances appearing in (7.11) can be evaluated from the table as needed.

Example 6.1 (continued, using (7.23) not (7.27)) The distance d between the two points in the plane is

$$d = \sqrt{r_1^2 + r_2^2 - 2r_1 r_2 \cos\phi},$$

where r_1 and r_2 are the lengths of the vectors from the origin to the two points and ϕ is the angle between these vectors. Estimating the distance of a point from the origin incurs a fixed relative error with standard deviation 0.2% and a random error with standard deviation 0.1 m. The angle is estimated independently.

The measurement error under the linear approximation is given in (7.8). The structural part of the error model is

$$\hat{r}_1 - r_1 = r_1 e_a + e_b, \qquad (9.1)$$
$$\hat{r}_2 - r_2 = r_2 e_a + e_c, \qquad (9.2)$$
$$\hat{\phi} - \phi = e_d. \qquad (9.3)$$

Table 9.1 shows one way in which this part of the model can be represented. The expression for the total error is placed in one column, while the expression for each component of error that affects more than one of the estimates is placed in a separate column of its own. Thus e_b, e_c and e_d feature only in the 'Total' column while e_a appears in that column and also enjoys a column of its own. More than one component might act on groups of estimates, so the general table will have more columns. This is indicated here by the heading 'From $\cdots$'. For example if an underlying component of error e_e had somehow affected all three estimates $\hat{r}_1$, $\hat{r}_2$, and $\hat{\phi}$ then there would be a column headed 'From e_e' with entries in every row.

The relevant statistical part of the model is

$$e_a \leftarrow D\left(0, 0.002^2\right),$$
$$e_b \leftarrow D(0, 0.1^2),$$
$$e_c \leftarrow D(0, 0.1^2),$$
$$e_d \leftarrow D(0, 2.50 \times 10^{-5}),$$

e_a, e_b, e_c, e_d have been drawn independently.

Replacing the entries in Table 9.1 by the corresponding variances gives Table 9.2, from which the covariance matrix can be calculated. The parent variances of the errors in the estimates of the three contributing quantities are

Table 9.1. Table of errors

Error	Total	From e_a	From $\cdots$
$\hat{r}_1 - r_1$	$r_1 e_a + e_b$	$+r_1 e_a$	
$\hat{r}_2 - r_2$	$r_2 e_a + e_c$	$+r_2 e_a$	
$\hat{\phi} - \phi$	e_d		

Table 9.2. Table of variances

Error	Total	From e_a	From $\cdots$	Units
$\hat{r}_1 - r_1$	≈ 0.022	≈ 0.012		m^2
$\hat{r}_2 - r_2$	≈ 0.034	≈ 0.024		m^2
$\hat{\phi} - \phi$	2.50×10^{-5}			rad^2

$$\mathrm{var_p}(\hat{r}_1 - r_1) = 0.002^2 \, r_1^2 + 0.1^2,$$
$$\mathrm{var_p}(\hat{r}_2 - r_2) = 0.002^2 \, r_2^2 + 0.1^2,$$
$$\mathrm{var_p}(\hat{\phi} - \phi) = 2.50 \times 10^{-5}$$

and the contributions to these total variances from the shared element e_a are $0.002^2 \, r_1^2$ and $0.002^2 \, r_2^2$. The approximate figures in Table 9.2 are obtained by replacing the unknowns r_1 and r_2 by their estimates $\hat{r}_1 = 54.8$ m and $\hat{r}_2 = 78.2$ m.

Thus Table 9.2 contains a single column of total parent variances for the contributing measured quantities and an additional column for the contribution from the shared element of error e_a. The parent covariance of any pair of errors can then be found from the data in this additional column. The covariance is the square root of the product of the entries in the two appropriate rows. Thus, the parent covariance of $\hat{r}_1 - r_1$ and $\hat{r}_2 - r_2$ is

$$\mathrm{cov_p}(\hat{r}_1 - r_1, \hat{r}_2 - r_2) \approx \sqrt{0.012 \times 0.024} = 0.017,$$

as we have seen in Table 7.1.

In more complicated problems there would be other columns associated with other shared elements of error. The parent covariance of any pair of errors is then found from the columns associated with the elements shared by these two errors. For each of these columns, the entries in the two appropriate rows are multiplied together and then the square root is taken. The covariance is the sum of these square roots over all these columns. For example suppose that measuring the distance of a point from the origin incurs an additional fixed error e_f taken to have arisen from a distribution with standard deviation 0.05 m. Then (9.1)–(9.2) would be replaced by the equations

$$\hat{r}_1 - r_1 = r_1 e_a + e_b + e_f, \tag{9.4}$$

$$\hat{r}_2 - r_2 = r_2 e_a + e_c + e_f \tag{9.5}$$

and Table 9.2 would contain an additional column with heading 'From e_f' with the value $0.05^2 = 0.0025$ in each of the first two rows. Then the parent covariance of $\hat{r}_1 - r_1$ and $\hat{r}_2 - r_2$ would be

$$\mathrm{cov}_\mathrm{p}(\hat{r}_1 - r_1, \hat{r}_2 - r_2) \approx \sqrt{0.012 \times 0.024} + \sqrt{0.0025 \times 0.0025},$$

so it would be larger by 0.0025. Also, the total parent variances for both $\hat{r}_1 - r_1$ and $\hat{r}_2 - r_2$ would be larger by 0.0025.

Table 9.2 might loosely be called a 'table of errors' or 'table of error variances'. Such a table is more fundamental than the covariance matrix because, in practice, the covariance matrix will be implicitly or explicitly constructed from such a table. The table could have various formats. The format described here is suitable for use in the next section, which presents a reformulation of the law of propagation of error expressed in terms of the entries in the table. Given the partial derivatives of $\mathcal{F}$, the variance of the overall measurement-error variate E can be calculated directly from the table.

9.2 An alternative propagation equation

The measurand equations at the beginning of Section 9.1 were linear, and so a linear analysis was exact. Now it must be acknowledged that we have abandoned the context of Chapter 8, where general non-linear functions were envisaged, and have returned to the domain of the linear approximation, where the law of propagation of error variance applies.

From (7.11), this law can be written as

$$\sigma^2(E) \approx \sum_{i=1}^{m} \left(\frac{\partial \mathcal{F}(\mathbf{x})}{\partial x_i}\right)^2 \sigma^2(E_i) + 2 \sum_{i=1}^{m-1} \sum_{j=i+1}^{m} \frac{\partial \mathcal{F}(\mathbf{x})}{\partial x_i} \frac{\partial \mathcal{F}(\mathbf{x})}{\partial x_j} \mathrm{cov}(E_i, E_j). \tag{9.6}$$

The inputs to this equation are the m partial derivatives, the m variances and the $m(m-1)/2$ (cross-)covariances. These (cross-)covariances, which describe dependency between errors in pairs of the m estimates, can only be calculated from some description of how the error variates interact. So not every input to (9.6) is a piece of raw information that is readily available. Therefore, an equivalent equation written in terms of the raw information would be useful.

It is reasonable to envisage the decomposition of the set of error sources into a set of independent elements acting additively. As in Example 6.1, where we encountered e_a, e_b, e_c and e_d as realizations of independent error variates E_a, E_b, E_c and

E_d, these elements (which sometimes have previously been referred to as 'components') will be identified using alphabetical rather than numerical subscripts. The general element will again be indicated using 'w'. So we can write

$$E_i = \sum_w \frac{\partial E_i}{\partial E_w} E_w.$$

Non-zero correlation between the variates E_i and E_j arises when one or more of the E_w elements is common to both. Let us define

$$\sigma(E_i; E_w) = \frac{\partial E_i}{\partial E_w} \sigma(E_w), \qquad (9.7)$$

so $\sigma^2(E_i; E_w)$ is the component of $\sigma^2(E_i)$ due to the element of error E_w. Then

$$\text{cov}(E_i, E_j) = \sum_w \sigma(E_i; E_w) \sigma(E_j; E_w)$$

with the typical term in the sum being zero because one or both of the factors $\sigma(E_i; E_w)$ and $\sigma(E_j; E_w)$ will be zero. In fact, we can also write

$$\text{cov}(E_i, E_j) = \sum_{w \in S} \sigma(E_i; E_w) \sigma(E_j; E_w),$$

where S denotes the set of elemental errors that contribute to more than one of the variates $E_1, \ldots, E_m$. The letter S stands for 'shared'. We see that non-zero covariances only arise from elements in S and that (9.6) can be written in the form (Willink, 2009)

$$\sigma^2(E) \approx \sum_{i=1}^{m} \left(\frac{\partial \mathcal{F}(\mathbf{x})}{\partial x_i} \right)^2 \sigma^2(E_i)$$

$$+ 2 \sum_{i=1}^{m-1} \sum_{j=i+1}^{m} \frac{\partial \mathcal{F}(\mathbf{x})}{\partial x_i} \frac{\partial \mathcal{F}(\mathbf{x})}{\partial x_j} \sum_{w \in S} \sigma(E_i; E_w) \sigma(E_j; E_w). \qquad (9.8)$$

Again, we might prefer to work in postmeasurement notation. Equations (9.7) and (9.8) become

$$\sigma_p(e_i; e_w) = \frac{\partial e_i}{\partial e_w} \sigma_p(e_w) \qquad (9.9)$$

and

$$\sigma_p^2(e) \approx \sum_{i=1}^{m} \left(\frac{\partial \mathcal{F}(\mathbf{x})}{\partial x_i} \right)^2 \sigma_p^2(e_i)$$

$$+ 2 \sum_{i=1}^{m-1} \sum_{j=i+1}^{m} \frac{\partial \mathcal{F}(\mathbf{x})}{\partial x_i} \frac{\partial \mathcal{F}(\mathbf{x})}{\partial x_j} \sum_{w \in S} \sigma_p(e_i; e_w) \sigma_p(e_j; e_w). \qquad (9.10)$$

9.2 An alternative propagation equation

A form of (9.10) that seems more suitable for computation is

$$\sigma_p^2(e) \approx \sum_{i=1}^{m} \left(\frac{\partial \mathcal{F}(\mathbf{x})}{\partial x_i}\right)^2 \left(\sigma_p^2(e_i) - \sum_{w \in S} \sigma_p(e_i; e_w)\sigma_p(e_j; e_w)\right)$$

$$+ \sum_{i=1}^{m} \sum_{j=1}^{m} \frac{\partial \mathcal{F}(\mathbf{x})}{\partial x_i} \frac{\partial \mathcal{F}(\mathbf{x})}{\partial x_j} \sum_{w \in S} \sigma_p(e_i; e_w)\sigma_p(e_j; e_w). \quad (9.11)$$

These equations show that the parent variance of the measurement error e can be expressed in terms of the partial derivatives, the parent variances $\{\sigma_p^2(e_i)\}$ and the contributions $\{\sigma_p^2(e_i; e_w)\}$ from the elements of error that contribute to more than one of the x_i estimates. No explicit $m \times m$ covariance matrix is required. Instead, we can use a table containing one column for the parent variances of the m quantity estimates and one column for each of the elemental errors that affect more than one of these estimates. We have just seen such a table in our analysis of Example 6.1; it is Table 9.2.

Example 6.1 (continued, using (7.23) not (7.27)) The distance d between two points is given by the measurand equation

$$d = \mathcal{F}(r_1, r_2, \phi) = \sqrt{r_1^2 + r_2^2 - 2r_1 r_2 \cos\phi}.$$

The estimate equation can be written as $\hat{d} = \mathcal{F}(\mathbf{x})$, where $\mathbf{x} = (x_1, x_2, x_3) = (\hat{r}_1, \hat{r}_2, \hat{\phi})$. The measurement error is

$$e \approx \frac{\partial \mathcal{F}(\mathbf{x})}{\partial x_1} e_1 + \frac{\partial \mathcal{F}(\mathbf{x})}{\partial x_2} e_2 + \frac{\partial \mathcal{F}(\mathbf{x})}{\partial x_3} e_3,$$

where

$$e_1 = \hat{r}_1 - r_1,$$
$$e_2 = \hat{r}_2 - r_2,$$
$$e_3 = \hat{\phi} - \phi.$$

The structural part of the error model is

$$e_1 = r_1 e_a + e_b,$$
$$e_2 = r_2 e_a + e_c,$$
$$e_3 = e_d.$$

Using the relevant statistical part of the error model, we then find the parent variances of e_1, e_2 and e_3 to be

$$\sigma_p^2(e_1) \approx 0.022 \text{ m}^2,$$
$$\sigma_p^2(e_2) \approx 0.034 \text{ m}^2,$$
$$\sigma_p^2(e_3) = 2.50 \times 10^{-5} \text{ rad}^2,$$

as before.

Only one elementary error affects more than one of the errors e_1, e_2 and e_3. This elementary error is e_a, so the set S contains one member, 'a'. We see that

$$\partial e_1/\partial e_a = r_1,$$
$$\partial e_2/\partial e_a = r_2,$$
$$\partial e_3/\partial e_a = 0,$$

so, from (9.9),

$$\sigma_p(e_1; e_a) = r_1 \sigma_p(e_a),$$
$$\sigma_p(e_2; e_a) = r_2 \sigma_p(e_a),$$
$$\sigma_p(e_3; e_a) = 0.$$

Replacing the unknowns r_1 and r_2 by their estimates $\hat{r}_1 = 54.8$ m and $\hat{r}_2 = 78.2$ m gives

$$\sigma_p^2(e_1; e_a) \approx 0.012 \text{ m}^2,$$
$$\sigma_p^2(e_2; e_a) \approx 0.024 \text{ m}^2,$$
$$\sigma_p^2(e_3; e_a) = 0.$$

We obtain Table 9.3, which is identical in content to Table 9.2.

In this example, $m = 3$ and (9.11) reduces to

$$\sigma_p^2(e) \approx \sum_{i=1}^{3} \left(\frac{\partial \mathcal{F}(\mathbf{x})}{\partial x_i}\right)^2 \left(\sigma_p^2(e_i) - \sigma_p(e_i; e_a)^2\right)$$
$$+ \sum_{i=1}^{3} \sum_{j=1}^{3} \frac{\partial \mathcal{F}(\mathbf{x})}{\partial x_i} \frac{\partial \mathcal{F}(\mathbf{x})}{\partial x_j} \sigma_p(e_i; e_a) \sigma_p(e_j; e_a). \quad (9.12)$$

The variance data required to implement this equation can be found from Table 9.3. The result is $\sigma_p^2(e) \approx 0.0781$ m^2, which agrees with (7.17) (because $e \approx e'$).

Now suppose there was the additional systematic error e_f with parent standard deviation 0.05 m affecting the estimates of r_1 and r_2, as in (9.4)–(9.5). The set S now contains a second member, 'f', for which $\partial e_1/\partial e_f = 1$, $\partial e_2/\partial e_f = 1$ and $\partial e_3/\partial e_f = 0$. So the figures 0.0025, 0.0025 and 0 appear in Table 9.3 in the rows of

Table 9.3. *Table of variances*

Index i	Error e_i	Parent variance $\sigma_p^2(e_i)$	From shared elements of error $\sigma_p^2(e_i; e_a)$	...	Units
1	$\hat{r}_1 - r_1$	≈ 0.022	≈ 0.012		m^2
2	$\hat{r}_2 - r_2$	≈ 0.034	≈ 0.024		m^2
3	$\hat{\phi} - \phi$	2.50×10^{-5}	0		rad^2

a column labelled $\sigma_\mathrm{p}^2(e_i; e_\mathrm{f})$, and the values of $\sigma_\mathrm{p}^2(e_1)$ and $\sigma_\mathrm{p}^2(e_2)$ are both increased by 0.0025. Equation (9.12) becomes

$$\sigma_\mathrm{p}^2(e) \approx \sum_{i=1}^{3} \left(\frac{\partial \mathcal{F}(\mathbf{x})}{\partial x_i}\right)^2 \left\{\sigma_\mathrm{p}^2(e_i) - \sigma_\mathrm{p}(e_i; e_\mathrm{a})^2 - \sigma_\mathrm{p}(e_i; e_\mathrm{f})^2\right\}$$
$$+ \sum_{i=1}^{3} \sum_{j=1}^{3} \frac{\partial \mathcal{F}(\mathbf{x})}{\partial x_i} \frac{\partial \mathcal{F}(\mathbf{x})}{\partial x_j} \left\{\sigma_\mathrm{p}(e_i; e_\mathrm{a}) \sigma_\mathrm{p}(e_j; e_\mathrm{a}) + \sigma_\mathrm{p}(e_i; e_\mathrm{f}) \sigma_\mathrm{p}(e_j; e_\mathrm{f})\right\}.$$

So uncertainty information can profitably be stored in a table of error variances. This table might be considerably smaller than an $m \times m$ covariance matrix. This economization is not seen in Example 6.1, where m is only 3, but it can be observed in the example given in the next section, where explicit storage of the table is practical but explicit storage of the matrix is not. Arguably, another advantage of the table is that it represents a more natural description of the measurement process.

9.3 Separate variances – measurement of functionals

Consider measuring many different points on a function, as in a typical calibration problem. The ideas discussed so far in this chapter are relevant because the results of these measurements will often be combined in order to estimate a functional, such as an integral of the function. (A *functional* is a function of a function. It is usually a scalar function, such as a norm, a derivative or an integral.) Suppose each point on the function is measured using the same technique, and that this technique incurs both a systematic error and a random error. Calculating the estimate of the integral will be simple, but calculating a corresponding figure of uncertainty will require separating the components of uncertainty that arise from fixed and random errors at each point. The following example illustrates these ideas (Willink, 2007c).

Example 9.1 A broadband radiometer is used to monitor the effective dose of erythemal (burn-causing) radiation received by persons exposed to sunlight. The contribution of sunlight to the dose is negligible at wavelengths less than 290 nm or greater than 400 nm. Calibration of the system involves measuring the dimensionless integral $\int_{290}^{400} v(\lambda) r(\lambda) d\lambda$, where λ is the wavelength in nanometres, $r(\lambda)$ is the response of the radiometer and $v(\lambda)$ is a weighting function representing the solar spectrum.

Data are available for the wavelengths 290 nm, 291 nm, ..., 400 nm, which we index by $i = 290, \ldots, 400$. Evaluation at 1 nm resolution is sufficient for sums to represent integrals, so the target value is the weighted sum

$$\theta \equiv \sum_{i=290}^{400} v_i r_i.$$

The data are the set of weights $\{v_{290}, \ldots, v_{400}\}$ and the set of response estimates $\{\hat{r}_{290}, \ldots, \hat{r}_{400}\}$. The measurement estimate is therefore $\hat{\theta} \equiv \sum_{i=290}^{400} v_i \hat{r}_i$. The weights can be taken as exact, so we write v_i, not $\hat{v}_i$. Table 9.4 shows a subset of the data for this problem, while Figure 9.1, which is based on actual data, shows potential data and the product $v_i \hat{r}_i$. From the full dataset, it is found that $\hat{\theta} = 1.935$.

We wish to obtain a 95% uncertainty interval for θ. The measurand θ is a weighted sum of quantities each measured using the same basic technique. If a correct analysis is to be carried out then the parent correlation between the $\hat{r}_i$ estimates must be taken into account. So the component of variance associated with systematic errors must be separated from the component associated with random errors. Several sources of systematic error affect the estimates at every wavelength: these sources of error may be treated as one. Also, owing to a small necessary change in technique midway through the frequency scan, there is an additional source of systematic error that affects only the estimates at the wavelengths 346–400 nm. Therefore, there are two elements of error that affect more than one of the $\hat{r}_i$ estimates, and this means that the set S has two elements. The corresponding data are given in the fourth and fifth columns of Table 9.4, where $e_i = \hat{r}_i - r_i$ and the two elements are indicated by 'e_a' and 'e_b'. In this example, the structure of the measurement is such that the shared components of error have slightly different sizes at each wavelength. So the effects of the elemental errors 'e_a' and 'e_b' are represented by standard deviations that change with wavelength. These happen to be relative standard deviations.

Table 9.4. *Calibration data for measuring exposure to erythemal radiation. The units for $\hat{r}_i$ are μW^{-1} nm cm^2 and the units for v_i are μW nm^{-1} cm^{-2}.*

i	$\hat{r}_i$	$\sigma_p(e_i)/r_i$	$\sigma_p(e_i; e_a)/r_i$	$\sigma_p(e_i; e_b)/r_i$	v_i
290	5.48×10^{-2}	0.0107	0.0107	0	2.89×10^{-4}
291	5.21×10^{-2}	0.0106	0.0106	0	2.81×10^{-4}
⋮	⋮	⋮	⋮	⋮	⋮
300	2.96×10^{-2}	0.0104	0.0104	0	6.25×10^{-1}
301	2.73×10^{-2}	0.0104	0.0104	0	1.18×10^{0}
⋮	⋮	⋮	⋮	⋮	⋮
330	2.19×10^{-4}	0.0155	0.0095	0	6.71×10^{1}
331	2.03×10^{-4}	0.0149	0.0092	0	5.96×10^{1}
⋮	⋮	⋮	⋮	⋮	⋮
345	8.97×10^{-5}	0.0295	0.0086	0	6.51×10^{1}
346	8.53×10^{-5}	0.0262	0.0086	0.0054	6.17×10^{1}
⋮	⋮	⋮	⋮	⋮	⋮
399	7.47×10^{-6}	0.3180	0.0080	0.0042	1.27×10^{2}
400	7.74×10^{-6}	0.3040	0.0064	0.0040	1.33×10^{2}

9.3 Separate variances – measurement of functionals

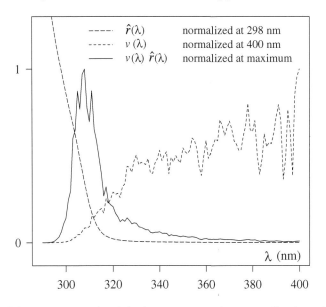

Figure 9.1 Response and weight data represented as normalized continuous functions of wavelength λ.

The measurement error $\hat{\theta} - \theta$ is

$$e = \sum_{i=290}^{400} v_i\, e_i.$$

So, from (9.11), which becomes an equality because the function is linear,

$$\sigma_p^2(e) = \sum_{i=290}^{400} v_i^2 \left(\sigma_p^2(e_i) - \sum_{w \in \{a,b\}} \sigma_p(e_i; e_w)^2 \right)$$
$$+ \sum_{i=290}^{400} \sum_{j=290}^{400} v_i v_j \sum_{w \in \{a,b\}} \sigma_p(e_i; e_w) \sigma_p(e_j; e_w).$$

Therefore,

$$\sigma_p^2(e) \approx \sum_{i=290}^{400} v_i^2 \hat{r}_i^2 \left(\left(\frac{\sigma_p(e_i)}{r_i} \right)^2 - \sum_{w \in \{a,b\}} \left(\frac{\sigma_p(e_i; e_w)}{r_i} \right)^2 \right)$$
$$+ \sum_{i=290}^{400} \sum_{j=290}^{400} v_i v_j \hat{r}_i \hat{r}_j \sum_{w \in \{a,b\}} \left(\frac{\sigma_p(e_i; e_w)}{r_i} \right) \left(\frac{\sigma_p(e_j; e_w)}{r_j} \right),$$

and this equation only involves terms found in Table 9.4. The parent standard deviation of $\hat{\theta}$ is then found from the full dataset to be $\sigma_p(e) = 0.026$ and the relative figure to be $\sigma_p(e)/\hat{\theta} = 1.3\%$.

In theory, the information forming Table 9.4 could have been stored in a 111×111 covariance matrix for direct application of the law of propagation of error (7.11). The use of (9.11) has removed the need for the explicit calculation of the elements of this large matrix.

Suppose that we had not recorded the $\sigma_p(e_i; e_a)/r_i$ and $\sigma_p(e_i; e_b)/r_i$ figures in Table 9.4, and that the only uncertainty data stored were the $\sigma_p(e_i)/r_i$ figures, as might have been the case if the radiometer had been calibrated without knowledge of the purpose to which it would later be put. On seeking to use the radiometer to measure θ or any other function of many r_i values, we would find ourselves unable to properly take into account the mixed nature of the errors. For example if we regard all the error as random, then the smoothing effect of summing over the spectrum is greater, and the standard uncertainty and fractional standard uncertainty of $\hat{\theta}$ have the smaller values of 0.006 and 0.3%. At the other extreme, if we take every component of error to affect the figure at each wavelength with perfect positive correlation then the figures rise to 0.037 and 1.9%. So there is a considerable amount of information that is provided by quoting the random and systematic components separately. Also, separate storage of the components of variance preserves some understanding about the process used to measure θ.

9.4 Worst-case errors – measurement in product testing

Measurement plays an important role in commerce through the inspection of manufactured products. Goods submitted by a 'producer' are tested to decide whether they are of sufficient quality to be put on the market – and often this test will be based on measurement. One standard situation is where a producer repeatedly sends items or batches of product for assessment using an established procedure, as is depicted in Figure 9.2. If the item or batch passes the inspection it is deemed suitable to be offered to the 'consumer'. Our discussion relates to this typical scenario (see Willink (2012a)).

If the entity under study (i.e. the item or batch) is to be shown fit for purpose then the assessment procedure must also be shown to be fit for purpose. This requires the acknowledgement of agreed standards of test performance by the producer and

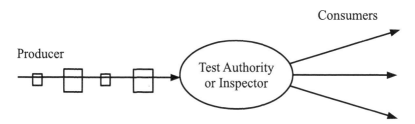

Figure 9.2 A typical scenario in the assessment of manufactured goods.

9.4 Worst-case errors – measurement in product testing

by the consumer, who is represented by the inspector. So the starting point for the development of any technique of product assessment must be an understanding of the legitimate expectations of both parties. The producer will consider the costs of various actions and decisions, and will think something like:

Based on my costs, I require at least 99% of the material I produce to reach the market. I cannot run a business unless this condition is met. I will construct my production system to achieve this, and the inspection procedure must be seen as part of this system.

Against this, an enforceable expectation on behalf of the consumer is that:

Only a small percentage of entities with quality below the required level will pass the inspection.

Thus, reliability for the producer relates to the long-run proportion of entities in his or her production that is deemed to conform, while reliability for the consumer relates to the long-run proportion of entities that have quality below the requirement but are deemed to conform.

Suppose that q is the true level of the quality characteristic of the entity, that D is the distribution of levels that the producer believes his or her manufacturing achieves and that q_0 is the minimum level required. The legitimate expectations of the parties can be expressed as follows. The test should be such that

$$\text{if } q \leftarrow D \text{ then Pr(entity will pass test)} \geq p \text{ and}$$
$$\text{if } q \leq q_0 \text{ then Pr(entity will pass test)} \leq \alpha.$$

The figure p is large, e.g. 99%, while the figure α is small. It follows that the producer must manufacture at a level above q_0.

Measurement error in this context

If we are to consider the proportion of material passing a test that is applied repeatedly then we must pay particular attention to errors that are constant. Imagine that the single quality characteristic for an item is its mass, that the minimum required mass is 250 g and that individual items are assessed using a technique with no random error but with an unknown fixed error δ. Consider the prospects for a certain competent producer who always manufactures items with mass 250.1 g. All the producer's items meet the requirement, and all give the same measurement result, $250.1 + \delta$. For any measurement result, there will necessarily be some specified critical level, x_{pass}, above which an item will be deemed to conform. So, depending on the relationship between x_{pass} and $250.1 + \delta$, either all of the producer's items will pass the test or none will pass the test. Yet the idea that the fixed error δ can be seen as having arisen from some continuous probability distribution, which is the idea put forward by the Working Group (Kaarls, 1980) and which has been adopted in earlier chapters of this book, suggests that the reliability of the test can be adjusted continuously by varying x_{pass}. This idea is seen to be inapplicable here.

In contrast, an alternative formulation is available. Designing a test procedure requires us (i) to consider the proportions of entities that pass this test and (ii) to acknowledge that this procedure will be continually used on goods from the same producer. So the errors that change from measurement to measurement must be treated differently from those that stay the same. A commonly used model is

$$x = q + \delta + e_{\text{ran}}, \qquad (9.13)$$

where q is the unknown value of the quality characteristic being measured, x is the result obtained, δ is an unknown amount of error affecting every measurement and e_{ran} is an unknown amount of error applying only to the measurement at hand. So δ is the total fixed error and e_{ran} is the total random error. The error e_{ran} will be treated as the outcome of a variate that takes a different value on each occasion, and we are entitled to propose an acceptable model for the distribution of this variate, say normal with mean zero and known variance σ_{ran}^2. However, in this context, as we have seen, we are not entitled to treat δ the same way. All we can do is propose realistic bounds on this error, say $\pm\delta_{\text{max}}$. The information describing the measurement uncertainty applicable in the test is then expressed by the random error variance σ_{ran}^2 and the bounds $\pm\delta_{\text{max}}$ on the fixed error.

Conformity assessment versus acceptance sampling

In order to see how this alternative formulation is to be applied, we must consider whether the uncertainty of measurement acts against the producer or against the consumer. The answer depends on whether the situation is one of *conformity assessment* or *acceptance sampling*.

Conformity assessment may be defined as (ISO, 2004, section 2.1):

demonstration that specified requirements relating to a product, process, system, person or body are fulfilled.

So, in conformity assessment, the focus is on showing beyond reasonable doubt that the entity under study meets its requirements, and the entity is not accepted unless the test shows this. However, in the field of acceptance sampling, a well-defined batch (or 'lot') of items is accepted unless the results obtained from a random sample show statistically that it fails to meet the requirements. This idea in acceptance sampling of 'accepted unless shown beyond reasonable doubt to fall short of the requirements' differs fundamentally from the idea in conformity assessment of 'not accepted unless shown beyond reasonable doubt to meet the requirements'. This difference has implications for the treatment of measurement uncertainty.

The presence of uncertainty must make the task of proving some claim more difficult. Therefore, in conformity assessment the presence of measurement uncertainty must act against the producer (who is trying to prove the adequacy of the product), while in acceptance sampling it must act against the consumer. Accordingly, in the context of a supplier–customer relationship, ISO 14253 (ISO, 1998, section 6.1) states

The principle behind the rules is the following: the uncertainty of measurement always counts against the party who is providing the proof of conformance or non-conformance ....

An assessment procedure for individual items

We now show how the figures of measurement uncertainty σ_{ran}^2 and $\pm\delta_{max}$ might feature in a conformity-assessment procedure that satisfies both the consumer's and producer's legitimate expectations. The situation assumed is simple: each item is assessed on the basis of a single measurement of a single quality characteristic, and the minimum required level of this characteristic is q_0. The test is to be such that:

1. no more than $100\alpha\%$ of items with quality below q_0 are to pass, and
2. at least $100p\%$ of items manufactured by the producer are to pass.

The proportion α will be small, while the proportion p will be high.

First, consider meeting the consumer's expectation. The item passes if the inspector concludes that $q \geq q_0$, and the probability of drawing this conclusion when it is incorrect is not to exceed α. So the null hypothesis being tested is $H : q < q_0$ and the allowable Type I error probability (or significance level[1]) is α. Suppose also that (9.13) is applicable, and that e_{ran} is the outcome of a variate with the distribution $N(0, \sigma_{ran}^2)$. The result x is the outcome of a corresponding variate X. After the item for submission has been selected, X has the distribution $N(q + \delta, \sigma_{ran}^2)$. So,

$$\Pr(X > q + \delta + z_{1-\alpha}\sigma_{ran}) = \alpha,$$

[1] In this book we make no distinction between a 'hypothesis test' and a 'significance test'.

where $z_{1-\alpha}$ is the $1-\alpha$ quantile of the standard normal distribution. (For example $z_{0.95} = 1.645$.) The unknown error δ is less than or equal to $\delta_{\max}$, so we know that

$$\Pr(X > q + \delta_{\max} + z_{1-\alpha}\sigma_{\text{ran}}) \leq \alpha.$$

It follows that H can be rejected with Type I error probability not exceeding α if

$$x > x_{\text{pass}}, \qquad x_{\text{pass}} \equiv q_0 + \delta_{\max} + z_{1-\alpha}\sigma_{\text{ran}}, \qquad (9.14)$$

in which x_{pass} is known. So the item passes the test of conformity if and only if (9.14) holds. This test satisfies the consumer's expectation.

Now consider meeting the producer's expectation. Suppose the production system generates items with quality characteristic having a normal distribution with fixed known variance σ_{prod}^2 and adjustable known mean q_{prod}. So the producer perceives q to be drawn from the distribution $\mathrm{N}(q_{\text{prod}}, \sigma_{\text{prod}}^2)$. Therefore, for the producer, the statistical model describing the whole process of selecting and testing an item is

$$X \sim \mathrm{N}\left(q_{\text{prod}} + \delta, \sigma_{\text{prod}}^2 + \sigma_{\text{ran}}^2\right).$$

The producer perceives the probability that this process results in a 'pass' to be

$$\Pr(X > x_{\text{pass}}) = \Phi\left(\frac{q_{\text{prod}} + \delta - x_{\text{pass}}}{\sqrt{\sigma_{\text{prod}}^2 + \sigma_{\text{ran}}^2}}\right),$$

where Φ is the standard-normal distribution function, e.g. $\Phi(1.645) \approx 0.95$. Because $\delta \geq -\delta_{\max}$, this probability is greater than or equal to the known figure

$$\Phi\left(\frac{q_{\text{prod}} - \delta_{\max} - x_{\text{pass}}}{\sqrt{\sigma_{\text{prod}}^2 + \sigma_{\text{ran}}^2}}\right). \qquad (9.15)$$

So the producer's expectation can be satisfied by setting q_{prod} high enough for the figure given by (9.15) to reach p.

Example 9.2 A producer sends items for conformity assessment. Each item is to have quality level at least 1000 in some units. Measurement of the quality level incurs a constant error with magnitude less than 2 and a normally distributed random error with mean 0 and standard deviation 1. So $q_0 = 1000$, $\delta_{\max} = 2$ and $\sigma_{\text{ran}} = 1$. The consumer has a right to expect that no more than 1% of items with true quality level less than 1000 will reach the market, so $\alpha = 0.01$. The test employed by the inspector will involve condition (9.14). We find that $x_{\text{pass}} = 1004.3$.

For the manufacturing process to be profitable, the producer requires at least 98% of the items to pass the conformity assessment, so $p = 0.98$. From production history, the producer knows that the quality levels of the items he or she produces have the normal distribution with standard deviation $\sigma_{prod} = 4$. So, from (9.15), the producer knows that the probability a random item passes the test is at least

$$\Phi\left(\frac{q_{prod} - 1006.3}{\sqrt{16+1}}\right).$$

This probability exceeds 0.98 when $q_{prod} \geq 1014.8$, so the producer's performance expectation will be satisfied if the mean quality level q_{prod} is set accordingly.

Summary

In conformity assessment or acceptance sampling, the legitimate expectations of the parties involved are to be respected. These expectations must relate to the proportions of items passing a test procedure that is applied repeatedly, which means that constant and random components of measurement error must be treated differently.

The basic idea for accommodating measurement uncertainty in this context is one of conservatism. In the analysis, worst-case values must be inserted for the parameters that describe the generation of the error in any measurement result. Thus (when a high measurement result is indicative of high product quality) the inspector must assume that each fixed error takes its maximum value in order to give the consumer the stated level of protection, while the producer must guard against the possibility that each fixed error takes its minimum value. For the producer who submits every product to the same inspection procedure, there is no comfort provided by the idea that corresponding systematic errors have different values in different procedures.

Thus, in this context of product testing, there is no logical basis for treating systematic errors as outcomes of random variables. A fixed component of error cannot be treated 'probabilistically', and the uncertainty associated with random and fixed components of error cannot be represented by a single standard deviation or 'standard uncertainty'. Similarly, the law of propagation of error variance (7.11) and the alternative propagation equation (9.8) are not applicable. These considerations imply that the understanding and treatment of measurement uncertainty proposed elsewhere in this book are not appropriate in product testing.

9.5 Time-scales for errors

The issues identified in this chapter present challenges to the idea that uncertainty in measurement is always properly represented by a single 'standard uncertainty'. Section 9.1 showed that knowing 'the standard uncertainty of x_1 for θ_1' and 'the standard uncertainty of x_2 for θ_2' is not necessarily sufficient to derive 'the standard uncertainty of $x_1 + x_2$ for $\theta_1 + \theta_2$'. Similarly, Section 9.3 demonstrated that the idea of a single standard uncertainty is inadequate in situations where many measurements made by the same method are combined to obtain an estimate of a derived quantity. And Section 9.4 showed this idea to be unhelpful when measurement forms the basis of a decision-making process intended to have a specified rate of accuracy. Thus, although in any measurement we might be able to calculate and state a 'standard uncertainty of measurement', we are not necessarily stating the most relevant quantity for the underlying purpose: the uncertainty associated with a measurement result should not be portrayed in a way that ignores how that result will subsequently be used.

The problem with reliance on a single 'standard uncertainty' is the failure to relate the time-scales of contributing errors to the purpose of the measurement. The idea that errors exist on different time-scales, which is by no means new (e.g. Croarkin (1989)), is seen in the basic experimental equation

$$x = \theta + e_{\text{ran}} + e_{\text{fix}}, \qquad (9.16)$$

which underlies the analyses described in this chapter. Here e_{ran} changes with complete randomness while maintaining a steady parent distribution but e_{fix} (or δ in Section 9.4) never changes after a measurement technique has been established. These are errors that arise from influences existing on extreme opposite time-scales.

An equation with greater generality would be

$$x = \theta + e_{\text{ran}} + e_{\text{mov}} + e_{\text{fix}},$$

where e_{mov} is a 'moving error' that exists on an intermediate time-scale, as described in Section 2.3. A full accommodation of moving errors would complicate the analysis. Thankfully, in many measurements the time-scale associated with the purpose of the measurement will be such that moving errors can be treated either as random or as fixed. For example when estimating the weighted sum of the spectral responses in Section 9.3, a moving error could be treated as a fixed error if it changed slowly compared with the time required to measure the responses over all the frequency range. We would then recover (9.16) as an appropriate model.

If doubt remains about how to treat errors that endure on an intermediate time-scale then we can fall back on the reasonable idea of preferring conservatism to unjustified optimism. To err by quoting an unnecessarily long interval (and increasing the success rate above the stated value) seems better than to err by reducing the success rate below the stated value (and quoting an unjustifiably short interval). Thus, in the problem of measuring the sum $\theta_1 + \theta_2$ in Section 9.1, in the integration problem of Section 9.3 and in the product-testing application of Section 9.4, a moving error should be treated as fixed, but in the problem of measuring the difference $\theta_1 - \theta_2$ in Section 9.1 it should be treated as random.

We may summarize this chapter by stating that different errors operate on different time-scales and that this fact is highly relevant to many measurement problems when we consider how the measurement result will be used. Our concepts of uncertainty evaluation must take this idea into account. The uncertainty information that is provided with a measurement estimate must be fit for purpose.

Part III

Related topics

Comparisons do ofttime great grievance.
John Lydgate
The Fall of Princes

There was always some ground of probability and likelihood mingled with his absurd behaviour.
Charles Dickens
Martin Chuzzlewit

What most experimenters take for granted before they begin their experiments is infinitely more interesting than any results to which their experiments lead.
Attributed to Norbert Wiener

A painting is never finished – it simply stops in interesting places.
Paul Gardner

10

Measurement of vectors and functions

Our focus has been on the estimation of quantities that are scalars. However, the measurand might be a p-dimensional vector with $p \geq 2$, or it might even be a function, for which we can regard p as infinite. So in this chapter we consider statistical techniques for studying more than one quantity at the same time, as in the fields of *multivariate analysis* and *simultaneous inference*. These techniques are relevant to the concept of measurement uncertainty that we have been discussing, the practical outworking of which is a specified minimum long-run success rate in an appropriate universe of problems. Here the measurand has p scalar components, each of which has its own target value, and success means bounding *every one* of these target values. Thus the ideas of success and uncertainty relate to the measurement as a single problem.

We describe the basic idea of a 'confidence region', which is the fundamental classical tool for estimation of a vector quantity. Subsequently, we describe the measurement of many scalars using 'simultaneous confidence intervals'. Also we consider statistical principles that are often overlooked when extending the ideas of inference to the multidimensional case: these principles relate to the efficiency and validity of our analyses. Then we turn our attention to the measurement of functions and to the more general idea of fitting a curve through a set of data points. We also briefly consider some ideas of calibration.

In Chapter 8 we could write $\boldsymbol{\theta} = (\theta_1, \ldots, \theta_m)$, for example, without concerning ourselves about whether this represented a row vector or a column vector. Now, however, some matrix algebra is involved, and all vectors in this chapter will be defined to be column vectors, as seems to be standard in multivariate statistics. Thus we now write $\boldsymbol{\theta} = (\theta_1, \ldots, \theta_p)'$ to indicate a column vector of p unknowns and $\mathbf{x} = (x_1, \ldots, x_p)'$ to indicate the estimate of $\boldsymbol{\theta}$. In keeping with common statistical notation, the letter p is used for the dimensionality of the problem.

Measurement of vectors and functions

10.1 Confidence regions

A confidence region is a generalization of a confidence interval to the situation where the quantity of interest is vector-valued. A confidence interval was defined in Definition 3.3. Analogously, a confidence region is defined as follows.

Definition 10.1 A *95% confidence region for a (target) vector* $\boldsymbol{\theta}$ *of p elements is a random region in p-dimensional space with probability 0.95 of enclosing* $\boldsymbol{\theta}$ *no matter what the values of all the unknown parameters.*

The region is random because its dimensions are random. The numerical region actually calculated in any problem will be called a *realized confidence region*. (So the region indicated on Figure 8.3 is a realized 95% confidence region for the column vector $(\theta_1, \theta_2)'$, for example.) The adjectives 'valid', 'approximate', 'average' and conditional that were introduced in the context of a confidence interval are applicable again.

The archetypal confidence region is the hyperellipsoidal region used to estimate the mean vector of a multivariate normal distribution. The theory is as follows.

Theory 10.1 Suppose $\boldsymbol{\theta} \equiv (\theta_1, \ldots, \theta_p)'$ is a column vector of p fixed unknowns. Let us prepare to estimate this vector on n independent occasions, and let X_{ij} be the variate for the jth estimate of θ_i. If the column vector $\mathbf{X}_j \equiv (X_{1j}, \ldots, X_{pj})'$ has the multivariate normal distribution with mean vector $\boldsymbol{\theta}$ and with a covariance matrix that does not depend on j then a 95% confidence region for $\boldsymbol{\theta}$ is the random set of all p-dimensional candidate vectors $\boldsymbol{\theta}_c$ for which

$$(\bar{\mathbf{X}} - \boldsymbol{\theta}_c)' \left(\frac{\mathbf{S}}{n}\right)^{-1} (\bar{\mathbf{X}} - \boldsymbol{\theta}_c) \leq T^2_{n-1,p,0.95}, \quad (10.1)$$

where $\bar{\mathbf{X}} = (1/n) \sum_{j=1}^n \mathbf{X}_j$ and

$$\mathbf{S} = \frac{1}{n-1} \sum_{j=1}^n (\mathbf{X}_j - \bar{\mathbf{X}})(\mathbf{X}_j - \bar{\mathbf{X}})',$$

with $T^2_{n-1,p,0.95}$ being the 0.95 quantile of the distribution of Hotelling's T^2 with parameters p and $n-1$ (Chatfield and Collins, 1980, pp. 116, 118). The region described by (10.1) is the interior of a random hyperellipsoid in p-dimensional space. The vector variate $\bar{\mathbf{X}}$ is the random sample mean vector and the $p \times p$ matrix variate $\mathbf{S}$ is the random sample covariance matrix. The quantity on the left of (10.1) is a random scalar and the relevant quantile of Hotelling's distribution is a constant scalar. Hotelling's distribution is related to the more-familiar F distribution by the equality

$$T^2_{n-1,p,0.95} = \frac{(n-1)p}{n-p} F_{p,n-p,0.95},$$

10.1 Confidence regions

where $F_{p,n-p,0.95}$ is the 0.95 quantile of the F distribution with p degrees of freedom in the numerator and $n - p$ degrees of freedom in the denominator.

The numerical data subsequently obtained form the vector $\bar{\mathbf{x}}$ and the matrix $\mathbf{s}$, which are the realizations of $\bar{\mathbf{X}}$ and $\mathbf{S}$. So a realized 95% confidence region for $\boldsymbol{\theta}$ is the set of all p-dimensional vectors $\boldsymbol{\theta}_c$ for which

$$(\bar{\mathbf{x}} - \boldsymbol{\theta}_c)' \left(\frac{\mathbf{s}}{n}\right)^{-1} (\bar{\mathbf{x}} - \boldsymbol{\theta}_c) \leq T^2_{n-1,p,0.95}.$$

These ideas are illustrated in the following example.

Example 10.1 Let V and I be the unknown voltage and current in a circuit in a steady state. Five independent measurements of both voltage and current are made simultaneously. The results obtained are shown in Table 10.1.

Each data pair can be regarded as having been drawn from a bivariate normal distribution with unknown mean vector $(V, I)'$ and unknown covariance matrix $\boldsymbol{\Sigma}$. Let us work in volts and milliamperes. Our estimate of the mean vector is

$$\begin{pmatrix} \hat{V} \\ \hat{I} \end{pmatrix} = \begin{pmatrix} 14.1982 \\ 30.5694 \end{pmatrix},$$

the elements of this vector being the sample means. Also, our estimate of the covariance matrix is found to be

$$\hat{\boldsymbol{\Sigma}} = \begin{bmatrix} 3.3070 \times 10^{-4} & -2.0339 \times 10^{-3} \\ -2.0339 \times 10^{-3} & 3.4473 \times 10^{-2} \end{bmatrix}.$$

(This vector and this matrix are $\bar{\mathbf{x}}$ and $\mathbf{s}$ in Theory 10.1.) In this case, $n = 5$ and $p = 2$, and the critical value on the right-hand side of (10.1) is 25.47. So a realized 95% confidence region for the target vector $(V, I)'$ is the set of all two-dimensional candidate vectors $(V_c, I_c)'$ for which

$$\begin{pmatrix} \hat{V} - V_c \\ \hat{I} - I_c \end{pmatrix}' \left(\frac{\hat{\boldsymbol{\Sigma}}}{5}\right)^{-1} \begin{pmatrix} \hat{V} - V_c \\ \hat{I} - I_c \end{pmatrix} \leq 25.47.$$

This region is the interior of the ellipse shown in Figure 10.1.

Table 10.1. *Recorded figures in five simultaneous measurements of the voltage and current in a circuit in steady state.*

i	$\hat{V}_i$ (V)	$\hat{I}_i$ (mA)
1	14.181	30.745
2	14.220	30.267
3	14.192	30.654
4	14.183	30.529
5	14.215	30.652

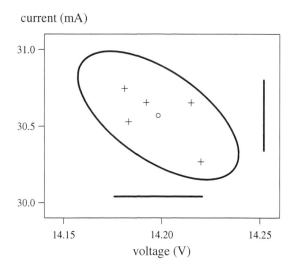

Figure 10.1 Elliptical realized 95% confidence region for the vector $(V, I)'$, and individual realized 95% confidence intervals for V and I.

A principle for maintaining statistical efficiency

Consider again Theory 10.1. The random hyperellipsoid has probability 0.95 of containing $\boldsymbol{\theta}$, so the random interval that extends between the limits of this hyperellipsoid in the ith dimension has probability at least 0.95 of containing θ_i. Thus, this random interval is a valid 95% confidence interval for θ_i. As shown in Appendix G, this interval is

$$\left[\bar{X}_i - \sqrt{T^2_{n-1,p,0.95}} S_i/\sqrt{n},\ \bar{X}_i + \sqrt{T^2_{n-1,p,0.95}} S_i/\sqrt{n}\right], \qquad (10.2)$$

where

$$\bar{X}_i = \frac{1}{n} \sum_{j=1}^{n} X_{ij}$$

and

$$S_i^2 = \frac{1}{n-1} \sum_{j=1}^{n} (X_{ij} - \bar{X}_i)^2.$$

Now $\sqrt{T^2_{n-1,1,0.95}} = t_{n-1,0.975}$ and $\sqrt{T^2_{n-1,p,0.95}} > t_{n-1,0.975}$ when $p > 1$, which is assumed here. Thus the interval (10.2) is longer than the interval

$$\left[\bar{X}_i - t_{n-1,0.975} S_i/\sqrt{n},\ \bar{X}_i + t_{n-1,0.975} S_i/\sqrt{n}\right], \qquad (10.3)$$

which is the 95% confidence interval that we would construct for θ_i from the same random data if θ_i had been the only parameter of interest. This reflects the fact that we are now making inference about p elements while maintaining the probability of drawing any false conclusion at the previous level of 0.05. More generally, $\sqrt{T^2_{n-1,p,0.95}} > \sqrt{T^2_{n-1,q,0.95}}$ for $p > q$. So the p-dimensional confidence region is larger in every dimension than the region applicable if we were only interested in estimating a subvector of $q < p$ elements.

Example 10.1 (continued) Suppose we had only been interested in estimating V. A realized 95% confidence interval for V would have been

$$\left[\hat{V} - t_{4,0.975}\sqrt{\hat{\Sigma}_{1,1}/5},\ \hat{V} + t_{4,0.975}\sqrt{\hat{\Sigma}_{1,1}/5}\right]$$

with $\hat{\Sigma}_{1,1} = 3.3070 \times 10^{-4}$ V^2. This is the interval [14.176 V, 14.221 V]. On the other hand, suppose we had only been interested in estimating I. A realized 95% confidence interval for I would have been

$$\left[\hat{I} - t_{4,0.975}\sqrt{\hat{\Sigma}_{2,2}/5},\ \hat{I} + t_{4,0.975}\sqrt{\hat{\Sigma}_{2,2}/5}\right],$$

with $\hat{\Sigma}_{2,2} = 3.4473 \times 10^{-2}$ mA^2. This is the interval [30.339 mA, 30.800 mA]. These intervals are marked on Figure 10.1. They are considerably shorter than the dimensions of the confidence region in voltage and current respectively.

So the extent of the confidence region in any dimension is greater than the length of the confidence interval that we would calculate for the corresponding individual element. This points us to a fundamental idea: statistical inference about any quantity of interest is strengthened by studying it alone. The idea of selecting 95% as the level of confidence means that we allow ourselves a probability of 0.05 of drawing an incorrect conclusion in our experiment, which is a 'doubt' of 5%. When we choose to study a vector of $p > 1$ parameters this doubt has to encompass more possibilities than when we study a single parameter, so the amount of doubt that we can afford to associate with any single parameter is reduced. It is therefore important to only formally estimate the quantities of genuine interest. Do not waste some of the allowable 5% failure probability on the estimation of other quantities!

10.2 Simultaneous confidence intervals

The idea of estimating one quantity with many components – for which we use a confidence region – differs in principle from the idea of estimating many unrelated scalars all at once – for which we would use a set of *simultaneous confidence intervals*.

Definition 10.2 A *set of simultaneous 95% confidence intervals* for fixed quantities $\theta_1, \ldots, \theta_p$ is a set of random intervals $\{I_1, \ldots, I_p\}$ such that

$$\Pr(I_1 \text{ will contain } \theta_1 \text{ and } \ldots \text{ and } I_p \text{ will contain } \theta_p) = 0.95$$

no matter what the values of all unknowns.

One approach to finding a valid set of simultaneous 95% confidence intervals for a set of quantities $\theta_1, \ldots, \theta_p$ is to construct a hyperellipsoidal 95% confidence region for the vector $\boldsymbol{\theta} = (\theta_1, \ldots, \theta_p)'$ and then to find a p-dimensional hyperrectangle that fully encloses this region. The variates for the maximum and minimum values in each dimension can be taken as the limits of the intervals. The hyperrectangle fully encloses the hyperellipsoid, so the success rate with this procedure will be at least 0.95, as required.

In the multivariate-normal case of Theory 10.1, this hyperrectangle has edges given by the intervals (Johnson and Wichern, 1998, Section 5.4)

$$\left[\bar{X}_i - \sqrt{T^2_{n-1,p,0.95} S_i^2/n}, \ \bar{X}_i + \sqrt{T^2_{n-1,p,0.95} S_i^2/n} \right], \qquad i = 1, \ldots, p, \tag{10.4}$$

as in (10.2). So (10.4) describes a set of valid simultaneous 95% confidence intervals for $\theta_1, \ldots, \theta_p$ (see Chatfield and Collins (1980), Section 7.3).

Example 10.1 (variation) Suppose that, instead of a 95% confidence region for $(V, I)'$, simultaneous 95% confidence intervals for V and I are required. The rectangle given by (10.4) fully encloses the ellipse given by (10.1). In this case, $p = 2$ and $n = 5$. So the pair of realized simultaneous intervals formed from the sides of this rectangle is

$$\left\{ \left[\hat{V} - \sqrt{T^2_{4,2,0.95} \hat{\Sigma}_{1,1}/5}, \ \hat{V} + \sqrt{T^2_{4,2,0.95} \hat{\Sigma}_{1,1}/5} \right], \right.$$
$$\left. \left[\hat{I} - \sqrt{T^2_{4,2,0.95} \hat{\Sigma}_{2,2}/5}, \ \hat{I} + \sqrt{T^2_{4,2,0.95} \hat{\Sigma}_{2,2}/5} \right] \right\}.$$

These two intervals are [14.158 V, 14.239 V] and [30.150 mA, 30.988 mA]. Figure 10.2 shows these intervals and the ellipse.

In general, the hyperrectangle constructed by this method will have considerably greater volume than the hyperellipsoid, so this will be a highly conservative procedure. A better procedure would have involved a 95% confidence region that was hyperrectangular in the first place. The Monte Carlo principle can be used to find such a confidence region when the joint error distribution is known.

Example 10.2 Simultaneous 95% confidence intervals are required for the output of a device at its three possible settings, 'low', 'medium' and 'high'. The output at

10.2 Simultaneous confidence intervals

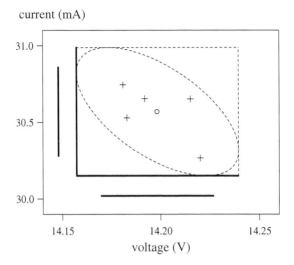

Figure 10.2 Realized simultaneous confidence intervals for V and I: (i) intervals formed from the sides of the rectangle bounding the confidence region and (ii) intervals calculated using the Bonferroni inequality.

each setting will be measured using a technique that incurs both a fixed error and a random error. The fixed error is known to lie between ± 1.4 units, with no value being more plausible than any other, and the random error will be drawn from the normal distribution with mean 0 and standard deviation 1.9 units.

The fixed and random errors are similar in scale, so neither can be disregarded. The fixed error will be deemed to have a uniform parent distribution, so the three overall measurement errors do not have a parent multivariate normal distribution. So we cannot appeal to the theory of the multivariate normal distribution in the construction of the intervals. However, the error distributions are fully known, and this facilitates a Monte Carlo analysis.

The three unknown outputs are θ_{low}, θ_{med} and θ_{high}. The measurement error variates E_{low}, E_{med} and E_{high} obey the model

$$E_{\text{sys}} \sim U(-1.4, 1.4),$$
$$E_{\text{low}} - E_{\text{sys}} \sim N(0, 1.9^2),$$
$$E_{\text{med}} - E_{\text{sys}} \sim N(0, 1.9^2),$$
$$E_{\text{high}} - E_{\text{sys}} \sim N(0, 1.9^2).$$

Therefore a simulated set of errors is found using the steps

1. $\tilde{e}_{\text{sys}} \leftarrow U(-1.4, 1.4)$,
2. $\tilde{e}_{\text{low}} \leftarrow N(\tilde{e}_{\text{sys}}, 1.9^2)$,
3. $\tilde{e}_{\text{med}} \leftarrow N(\tilde{e}_{\text{sys}}, 1.9^2)$,
4. $\tilde{e}_{\text{high}} \leftarrow N(\tilde{e}_{\text{sys}}, 1.9^2)$.

Repetition of this simulation enables us to find the value a such that in 95% of simulations $|\tilde{e}_{\text{low}}|$, $|\tilde{e}_{\text{med}}|$ and $|\tilde{e}_{\text{high}}|$ are all less than or equal to a.

We find that $a = 4.91$. Therefore, the random intervals $[X_{\text{low}} - 4.91, X_{\text{low}} + 4.91]$, $[X_{\text{med}} - 4.91, X_{\text{med}} + 4.91]$ and $[X_{\text{high}} - 4.91, X_{\text{high}} + 4.91]$ are a set of simultaneous 95% confidence intervals for θ_{low}, θ_{med} and θ_{high}. The confidence coefficient of this procedure does not exceed 0.95 because a has been chosen accordingly. So this procedure is more efficient than the process of calculating a curved 95% confidence region for $\boldsymbol{\theta} = (\theta_{\text{low}}, \theta_{\text{med}}, \theta_{\text{high}})'$ and subsequently finding a cuboid bounding this region.

The actual measurement results are $x_{\text{low}} = 45.2$, $x_{\text{med}} = 59.3$ and $x_{\text{high}} = 73.6$. So a set of realized simultaneous 95% confidence intervals for θ_{low}, θ_{med} and θ_{high} are $[40.3, 50.1]$, $[54.4, 64.2]$ and $[68.7, 78.5]$.

The idea of simultaneous confidence intervals is an important idea in the field of *simultaneous inference* (e.g. Miller (1980)), where a single study or experiment is used to draw conclusions on more than one question. When conducting many hypothesis tests simultaneously, the term *multiple comparisons* is often used.

The principal Bonferroni inequality

The probability that at least one of p events will occur cannot exceed the sum of the probabilities of the individual events. That is, for every set of p events $\{B_1, \ldots, B_p\}$,

$$\Pr(B_1 \text{ or } \ldots \text{ or } B_p) \leq \sum_{i=1}^{p} \Pr(B_i). \tag{10.5}$$

Let A_i be the complement of B_i. Rewriting (10.5) leads to

$$\Pr(A_1 \text{ and } \ldots \text{ and } A_p) \geq 1 - \sum_{i=1}^{p} \Pr(B_i), \tag{10.6}$$

with $\Pr(B_i) = 1 - \Pr(A_i)$. Suppose that the quantities $\theta_1, \ldots, \theta_p$ are to be measured simultaneously and that A_i is the event that the confidence interval for θ_i is successful. Then (10.6) is

$$\Pr(\text{all will be successful}) \geq 1 - \sum_{i=1}^{p} \Pr\left(\begin{array}{c}\text{the } i\text{th confidence interval}\\ \text{will not be successful}\end{array}\right).$$

This inequality shows how confidence levels in the estimation of p individual quantities can be chosen so that the joint probability of every interval being successful is kept at or above 95%. It enables us to state the following result.

10.2 Simultaneous confidence intervals

Result 10.1 Let I_i be a $(1 - 0.05/p) \times 100\%$ confidence interval for a measurand θ_i. The set of intervals $\{I_1, \ldots, I_p\}$ is a set of valid simultaneous 95% confidence intervals for the measurands $\theta_1, \ldots, \theta_p$.

Inequality (10.5)–(10.6) is the simplest member of the family known as the Bonferroni inequalities (Feller, 1968, p.110). It is often referred to as *the* Bonferroni inequality.

Example 10.1 (variation, continued) Suppose that simultaneous 95% confidence intervals for V and I are required. Result 10.1 with $p = 2$ implies that individual 97.5% confidence intervals for V and I form a pair of simultaneous 95% confidence intervals for V and I. So a pair of realized simultaneous 95% confidence intervals for V and I is

$$\left\{ \left[\hat{V} - t_{4,0.9875}\sqrt{\hat{\Sigma}_{1,1}/5},\ \hat{V} + t_{4,0.9875}\sqrt{\hat{\Sigma}_{1,1}/5} \right], \right.$$
$$\left. \left[\hat{I} - t_{4,0.9875}\sqrt{\hat{\Sigma}_{2,2}/5},\ \hat{I} + t_{4,0.9875}\sqrt{\hat{\Sigma}_{2,2}/5} \right] \right\}.$$

These are the intervals [14.170 V, 14.227 V] and [30.279 mA, 30.860 mA], which are marked on Figure 10.2. They are considerably shorter than the intervals found from the rectangle bounding the elliptical confidence region.

The Bonferroni inequality is particularly useful when we do not know the dependence structure between components of error.

Example 10.2 (variation) Suppose it cannot be assumed that the error drawn from the uniform distribution was the same for the three settings, 'low', 'medium' and 'high'. The previous analysis becomes invalid. However, putting $p = 3$ in Result 10.2 shows that a valid set of simultaneous 95% confidence intervals for the three outputs can be obtained by forming individual 98.34% confidence intervals for the outputs at each of the three settings. Consider the measurement at the 'low' setting. A simulated error is found using the steps

1. $\tilde{e}_{\text{sys}} \leftarrow U(-1.4, 1.4)$,
2. $\tilde{e}_{\text{low}} \leftarrow N(\tilde{e}_{\text{sys}}, 1.9^2)$.

Then the value a is found such that in 98.34% of simulations $|\tilde{e}_{\text{low}}|$ is less than or equal to a. This value is found to be 5.69, so the interval $[X_{\text{low}} - 5.69, X_{\text{low}} + 5.69]$ is a 98.34% confidence interval for θ_{low}. The same value of a is applicable with θ_{med} and θ_{high}.

The actual measurement results are $x_{\text{low}} = 45.2$, $x_{\text{med}} = 59.3$ and $x_{\text{high}} = 73.6$, so realized simultaneous 95% confidence intervals for $\theta_{\text{low}}, \theta_{\text{med}}$ and θ_{high} are [39.5, 50.9], [53.6, 65.0] and [67.9, 79.3]. These intervals are longer than those calculated when we knew the dependence structure of the errors.

The level of conservatism that arises from the use of the Bonferroni inequality is small when the relevant variates are independent. For example, suppose a set of $p = 5$ simultaneous 95% confidence intervals is required. By Result 10.1, each individual confidence level can be set to $(1 - 0.05/p) \times 100\% = 99\%$. If the five events of success happen to be independent then the actual probability that all the intervals will be successful is $0.99^5 \approx 0.951$, so the degree of conservatism is small.

Result 10.1 implies that reducing p allows the confidence levels for the individual intervals to be increased, which leads to a shortening of these intervals. So, as has already been stated, inference is strengthened by studying only the quantities of genuine interest.

10.3 Data snooping and cherry picking

We have observed that unnecessary simultaneous inference leads to statistical inefficiency, which – in our context – is the quotation of intervals that are longer than necessary. However, there is a converse mistake that can be made, and this mistake is much more serious. It is the mistake of failing to acknowledge that when we commenced our study we had many questions in mind: it is the mistake of failing to recognize that we were engaged in simultaneous inference in the first place. If we do not make an adjustment like that implied by the Bonferroni inequality then the resulting intervals become the outputs of incorrect, invalid, procedures.

Two relevant phrases used colloquially are 'cherry picking' and 'data snooping'. If we conduct many hypothesis tests and claim statistical significance for all those tests with p values less than 0.05 without acknowledging the lack of significance in the other tests then we are choosing to pick only the most ripe cherries from the tree. Similarly if we only decide to apply a test after informally observing a suspicious pattern in the data then we are 'snooping', which is sneaking a peak before addressing the dataset in a proper fashion. Neither of these practices meets the standards of statistical logic. Forming a confidence interval is analogous to inverting many hypothesis tests, so the practices of cherry picking and data snooping in multiple hypothesis testing have counterparts in the formation of simultaneous confidence intervals. The proportion of experiments in which one or more of the p intervals will be unsuccessful will exceed 0.05 unless these practices are avoided.

Cherry picking, data snooping and the inflated failure rate that follows are not phenomena confined to the frequentist approach to statistics. They are also relevant in the Bayesian paradigm of statistics, although there they are somewhat disguised. So the Bonferroni inequality for the adjustment of probability levels is also applicable in Bayesian analyses.

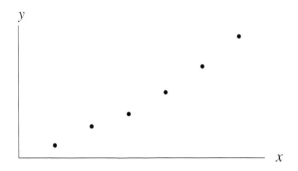

Figure 10.3 A set of (x, y) data.

10.4 Curve fitting and function estimation

Let us now consider the idea of fitting a smooth curve to a scattered set of two-dimensional data $\{(x_1, y_1), \ldots, (x_n, y_n)\}$, such as the set of data indicated in Figure 10.3. I have written 'idea' rather than 'task' because often a technique of fitting seems chosen with little obvious regard for the purpose: the objective has not been made clear. There are at least three possibilities.

1. No unique function is thought to underlie the data, but we seek some function to act as a representative summary of them.
2. There is a unique function giving rise to the data and we wish to assess the truth of a hypothesis that this function has a certain form.
3. There is a unique function $f(x)$ giving rise to the data and we wish to estimate this function for some reason associated with future values of x or y.

These broad objectives correspond to three possible purposes of 'modelling', namely:

1. to summarize a dataset,
2. to investigate the science of some system – an advancement being made if a scientific hypothesis or 'model' is shown to be incompatible with a set of experimental data,
3. to form predictions about the response of a system at future values of input variables.

The first of these tasks, which is the idea of modelling to summarize a dataset, seemed important to a referee of one of my papers (Willink, 2008). A mild paraphrase of some of the referee's comments is as follows:

I would not so much insist on the view that there is a true function to be approximated or estimated. We never know whether such a true relation between quantities exists, and if so what it looks like. We do not even know whether the concept of a straight line has a counterpart in reality. Our experience shows that nearly every theory that we have set up is wrong. Nowadays, we are able to approximate a given series of points on a plane as closely as we wish. But we are not interested in doing so. Instead, we look for an effective and useful description of the dataset.

Presumably, some readers will agree with those comments. However, the idea of summarizing or describing a dataset needs further explanation. The dataset is its own perfect summary, and it might be written down faster than the computer used to fit the function will boot – so why summarize the data at all? And what properties of the data should make up the summary? How well should each point be represented? The task of 'describing a dataset' seems empty without a clearer purpose being identified. And if we actually believe that there is no true function underlying the data then why do we usually form such a summary using regression techniques, which are based in logic on the idea that such a function – the regression function – exists?

The second of these concepts, which is the idea that data-based modelling can be used to investigate the science underlying a system, is outside the scope of this book.

So our focus will be on the third of these concepts of modelling, which is the idea of measuring the response of a system in order to predict future values of y from observations of x, or vice versa. The appropriate analysis depends greatly on what is known about the data points and the function $f(x)$. Again, there are different possibilities.

3.1 The data points $\{(x_i, y_i)\}$ are known to have negligible error, so the function $f(x)$ can be seen as passing through each of these points. In this case either (i) we seek an estimating function $\hat{f}(x)$ that also passes through each of the points or (ii) we permit $\hat{f}(x)$ to miss the points in order to obtain an approximation with a suitably simple form. (See Figure 10.4.)

3.2 The uncertainty associated with the x or y co-ordinates cannot be ignored, so we conclude that $f(x)$ does not pass through the data points. Hence we permit the estimating function $\hat{f}(x)$ to miss the data points. Often the form of $f(x)$, say linear, is suggested from physical considerations and our estimate $\hat{f}(x)$ is to have the same form.

The usual situation is described in point 3.2: some error will exist in x_i or y_i or both. In this case, our goal might be:

3.2.1 to estimate $f(x)$ at a known value of x, or to estimate x for a known value of $f(x)$,

3.2.2 to estimate $f(x)$ at some value of x for which we only have an estimate, or to estimate x at some value of $f(x)$ for which we only have an estimate,

3.2.3 to predict a measurement result rather than the target co-ordinate on $f(x)$, for example, to predict y_{n+1} given x_{n+1}.

10.4 Curve fitting and function estimation

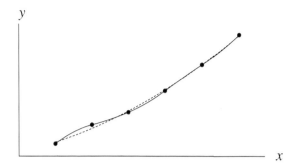

Figure 10.4 Two functions fitted to the data of Figure 10.3. Solid line: a quadratic spline passing through all the points. Dashed line: the least-squares quadratic function, which does not actually pass through any of the points.

These are different statistical problems, and so considerable thought should go into choosing or developing the technique to be employed.

Often the estimate of the function will be used many times. This complicates the question of evaluating and expressing the relevant measurement uncertainty or prediction uncertainty. Suppose that we wish to estimate $f(x)$ by some function $\hat{f}(x)$ and to provide a 95% uncertainty statement with that estimate. Are we envisaging the subsequent use of $\hat{f}$ at a single value of x such as a fixed operating point, say $\acute{x}$? Or are we imagining the subsequent use of $\hat{f}$ at many points in some known interval $[x_{\min}, x_{\max}]$, in which case concepts of simultaneous inference are relevant? The nature of the statement of uncertainty depends on our answer. Are we providing (i) a realized 95% confidence interval for $f(\acute{x})$ or (ii) a realized 95% *confidence band* for $f(x)$ for all x in $[x_{\min}, x_{\max}]$, i.e. a confidence region for the infinite set of $f(x)$ values that form the function in $[x_{\min}, x_{\max}]$? In accordance with the principle noted earlier in the chapter, the width of this confidence band at $x = \acute{x}$ will be greater than the width of the confidence interval.

Example 10.3 (Least-squares linear regression) Suppose we possess data $(x_1, y_1), \ldots, (x_n, y_n)$ satisfying $y_i = f(x_i) + e_i$, where $f(x)$ is an unknown linear function. If the x_i values were predetermined and the e_i values were independently drawn from a shared normal distribution with mean zero then the full theory of least-squares linear regression is applicable. Summation without limits is over $i = 1, \ldots, n$. Let us define

$$\bar{x} = \frac{\sum x_i}{n},$$

$$\bar{y} = \frac{\sum y_i}{n},$$

$$b = \frac{\sum x_i y_i - n\bar{x}\bar{y}}{\sum x_i^2 - n\bar{x}^2},$$

$$a = \bar{y} - b\bar{x},$$

$$s^2 = \frac{\sum(y_i - a - bx_i)^2}{n-2},$$

$$v = \frac{\sum(x_i - \bar{x})^2}{n}.$$

An unbiased point estimate of $f(\acute{x})$ for any specific $\acute{x}$ is the figure $a + b\acute{x}$. More importantly, under the conditions stated, a realization of a $100P\%$ confidence interval for $f(\acute{x})$ is the interval with limits (e.g. Miller (1986), p. 176)

$$a + b\acute{x} \pm t_{n-2,(1+P)/2} \frac{s}{\sqrt{n}} \sqrt{1 + \frac{(\acute{x} - \bar{x})^2}{v}}. \tag{10.7}$$

Also, under these conditions a realization of a $100P\%$ confidence region for $f(x)$ that holds simultaneously for all x has bounding curves given by the hyperbolae

$$a + bx \pm \sqrt{2F_{2,n-2,P}} \frac{s}{\sqrt{n}} \sqrt{1 + \frac{(x - \bar{x})^2}{v}} \tag{10.8}$$

where $F_{2,n-2,P}$ is the P quantile of the F distribution with 2 and $n-2$ degrees of freedom (e.g. Miller (1986), p. 176). These curves define a realized *confidence band*. The distance between these curves at the point $\acute{x}$ is longer than the length of the interval given in (10.7). For example when $P = 0.95$ and $n = 10$, $\sqrt{(2F_{2,n-2,P})} = 2.99$ but $t_{n-2,(1+P)/2} = 2.31$. (See also Mandel (1964).)

The line $a+bx$ and the band defined by the curves (10.8) for a certain situation are shown in Figure 10.5. The hyperbolic form of these curves might be inconvenient. Graybill and Bowden (1967) and Dunn (1968) have given curves with the 'bow-tie' forms also illustrated in Figure 10.5. Dunn's construction involves the idea that, in

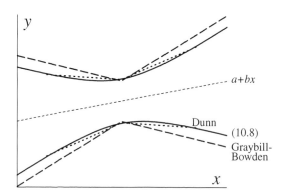

Figure 10.5 Confidence bands for the estimation of a straight line with a certain dataset. Solid curves: band defined by the hyperbolic curves of (10.8). Dashed lines: band of Graybill and Bowden (1967). Dotted line segments: band of Dunn (1968).

practice, only a finite range of x values will be of interest. Willink (2008) gives more details.

Furthermore, under the conditions stated, if a measurement is subsequently to be made at $\acute{x}$ then – before the generation of the data $y_1, \ldots, y_n$ and the figures a and b – there is probability P that the result will lie in the interval (Acton, 1959, p. 36; Lindgren, 1968, p. 455)

$$\check{a} + \check{b}\acute{x} \pm t_{n-2,(1+P)/2} \frac{\check{s}}{\sqrt{n}} \sqrt{1 + n + \frac{(\acute{x} - \bar{x})^2}{v}},$$

where $\check{a}$, $\check{b}$ and $\check{s}$ are the variates for a, b and s. This type of interval is called a *prediction interval*.

If the conditions are satisfied only approximately then the confidence interval, confidence region and the prediction interval are approximate also.

So there are many possible reasons for fitting a curve to data and – when an underlying function is being measured – statements of measurement uncertainty can be of at least two types. Obviously, the purpose of the analysis should influence the choice of method and the type of uncertainty statement. The starting point must be a clear definition of the problem.

10.5 Calibration

Our attention now turns to the subject of calibration, which is of great importance in measurement science. Arguably, the basic idea of calibration is function estimation, which has been discussed to some extent in Section 10.4. However, the subject of calibration – and especially the idea of uncertainty in calibration – deserves a much fuller treatment, and this should be a treatment given by someone with more practical experience and a broader understanding than the present author. So our attention here is restricted to a few underlying concepts.

The definition of calibration given in the *International Vocabulary of Metrology* (VIM, 2012, clause 2.39) may be summarized as follows

Calibration is an operation that establishes a relation for obtaining a measurement result from an indication.

In the statistical literature, the 'calibration' problem presupposes the existence of a unique underlying function $f(x)$ on some interval and generally involves the requirements of the usual parametric regression analysis, which are that $f(x)$ has a known form, that there is no error in the x_i values, that the errors $\{y_i - f(x_i)\}$ are independently drawn from some distribution with a specified form and that the size of the dataset was determined without being influenced by the values of the data. In that context, various problems and solutions have been put forward – few of which seem to relate to the task of calibration in a practical measurement setting. But just as the statistician must realize that the practical context has harsh realities, so the measurement scientist must understand that often there are statistical assumptions underlying the analysis technique that he or she employs. Both types of scientist are in danger of oversimplifying the problem, one to give logically meaningful results with little practical applicability and the other to give results that seem to address a practical problem but which actually rely on logic that is inapplicable.

As can be inferred from these comments, I believe that 'calibration' can mean different things to different people. It would be presumptuous of me to suggest that I have any real grasp of the breadth of calibration problems encountered in practice. So the remainder of this section contains only an illustration of a conceptual difference between two calibration situations.

Calibration of a measuring device and a generating device

Consider a device that is designed to measure a stimulus. The calibration of such a device usually involves presenting it with a set of known stimuli $\{\theta_i\}$, recording the corresponding set of indications $\{x_i\}$ and finding a function $\hat{g}$ such that $\theta_i \approx \hat{g}(x_i)$ for each i. (The notation $\hat{g}$ suggests that we see this function as an estimate of an unknown target function g.) When the device is subsequently used, an unknown stimulus θ is presented to the device, the indication x is recorded and the value of the stimulus is estimated by $\hat{\theta} = \hat{g}(x)$. This process can be represented as:

$$\begin{array}{lccc} \text{calibration} & \{\theta_i\}, \{x_i\} & \longrightarrow & \hat{g}, \\ \text{time of use} & x & \stackrel{\hat{g}}{\longrightarrow} & \hat{\theta}. \end{array}$$

The calibration is carried out so that the error $\hat{\theta} - \theta$ will be small in magnitude.

In contrast, consider a device intended to produce a known signal, for example radiation of a certain wavelength. The calibration of this device might involve relating a set of known signals $\{\theta_i\}$ to the set of indications $\{x_i\}$ that appear on the device when it generates matching signals, and then finding a function $\hat{h}$ such that $x_i \approx \hat{h}(\theta_i)$ for each i. (This contrasts with the earlier idea of expressing θ_i as a function of x_i.) When the device comes to be used and a signal θ is required, the device is set to the indication $x = \hat{h}(\theta)$. This process can be represented as:

$$\begin{array}{lccc} \text{calibration} & \{\theta_i\}, \{x_i\} & \longrightarrow & \hat{h}, \\ \text{time of use} & \theta & \stackrel{\hat{h}}{\longrightarrow} & x. \end{array}$$

The calibration is carried out so that the signal generated will be as close as possible to the signal intended, θ.

10.5 Calibration

The two processes are different. So the calibration of a 'measuring device' is to be distinguished from the calibration of a 'generating device'. The two tasks differ conceptually as a consequence of the devices having different roles. This observation supports the basic claim being made in this section: the idea of 'calibration' can differ from situation to situation. As a consequence, there will be differences between the appropriate concepts of calibration uncertainty. Moreover, these concepts of uncertainty will depend on whether or not the functions $\hat{g}$ and $\hat{h}$ are to be understood as estimates of unknown target functions g and h, which is suggested by our choice of notation.

11
Why take part in a measurement comparison?

Our foray into multivariate problems is over, and we return to the idea of measuring a scalar quantity θ. The result of the measurement is a known scalar x, and associated with this result will be estimates $\hat{\sigma}_A^2$ and $\hat{\sigma}_B^2$ of the parent variances of the total Type A error and the total Type B error. The validity of the figure $\hat{\sigma}_A^2$ can be demonstrated if the experimental data are recorded, but the figure $\hat{\sigma}_B^2$ will often have been formed by expert judgement and might be questioned. In this chapter we describe how the accuracy of $\hat{\sigma}_B^2$ can be examined using data obtained when other laboratories measure θ. These data might also provide an estimate of the value of the fixed component of error, which is interpretable as a 'laboratory bias' or 'experimental bias'. So there are at least two reasons for participating in an inter-laboratory comparison.

The type of comparison studied is the simplest possible: a number of laboratories independently measure θ, each providing an estimate with corresponding uncertainty information. We describe how such a comparison can be used to examine the accuracy of $\hat{\sigma}_B^2$ as an estimate of σ_B^2 and then we give a critical discussion of the statistical method involved, which is a 'goodness-of-fit' test. Subsequently, we focus on the estimation of the actual total Type B error, and we show that the comparison provides much more evidence about the value of this error than about its parent variance σ_B^2. Then we discuss the estimation of bias in the measurement technique and the corresponding application of a correction in future measurements.

11.1 Examining the uncertainty statement

In almost all cases we are able to regard x as the outcome of a variate with mean equal to the target value θ and with variance being the sum of a component σ_A^2 estimated from statistical data and a component σ_B^2 associated with errors evaluated by other means. The model can then be written as

11.1 Examining the uncertainty statement

$$x = \theta + e_A + e_B,$$
$$e_A \leftarrow D(0, \sigma_A^2),$$
$$e_B \leftarrow D(0, \sigma_B^2),$$

where e_A is the total Type A error and e_B is the total Type B error. The estimates of σ_A^2 and σ_B^2 are $\hat{\sigma}_A^2$ and $\hat{\sigma}_B^2$. This can be summarized as

$$x = \theta + e,$$
$$e \leftarrow D(0, \sigma^2),$$
$$\sigma^2 = \sigma_A^2 + \sigma_B^2$$

and the estimate of the total variance σ^2 is

$$\hat{\sigma}^2 = \hat{\sigma}_A^2 + \hat{\sigma}_B^2.$$

Suppose that n other laboratories independently measure θ and that the kth of these laboratories provides an estimate x_k of θ and an estimate $\hat{\sigma}_k^2$ of the relevant total variance. The result obtained is described by

$$x_k = \theta + e_k, \qquad k = 1, \ldots, n,$$
$$e_k \leftarrow D(0, \sigma_k^2)$$

with $\hat{\sigma}_k^2$ being the estimate of σ_k^2. These models and the data can be used to examine the accuracy of $\hat{\sigma}^2$ as an estimate of σ^2. Because we have accepted the accuracy of $\hat{\sigma}_A^2$, this implies that we are examining the accuracy of $\hat{\sigma}_B^2$ as an estimate of σ_B^2.

Ideally, we would like this process to confirm that $\hat{\sigma}_B^2$ is an approximation to σ_B^2 that is fit for purpose; this purpose being to carry out measurements with at least the advertised success rate while generating short intervals. If $\hat{\sigma}_B^2$ is low compared with σ_B^2 then measurements made using our technique will have a lower success rate, while if $\hat{\sigma}_B^2$ is high compared with σ_B^2 then the intervals will be unnecessarily long. So we wish to find if σ_B^2 differs from $\hat{\sigma}_B^2$ by a material amount. Regrettably, as we shall see, the comparison often cannot answer this question!

In many situations, the task will be approached by a statistical test of the hypothesis

$$H : \sigma^2 = \hat{\sigma}^2.$$

In carrying out this test, we are thinking or behaving as follows. If our measurement estimate x is too far from some figure that represents the location of the x_k values then we have evidence that $\sigma^2 > \hat{\sigma}^2$, while if x is too close to that figure then we have evidence that $\sigma^2 < \hat{\sigma}^2$. In either case, the idea that the data show σ^2 to differ from $\hat{\sigma}^2$ is (somewhat illogically) taken as evidence that σ^2 differs from $\hat{\sigma}^2$ by an amount that is troublesome. In contrast, if x is sufficiently close to the figure that represents the location of the x_k values then (somewhat illogically) we proceed as if σ^2 were equal to $\hat{\sigma}^2$.

It is helpful to now use premeasurement notation, which involves variates rather than realized values. The models above become

$$\left.\begin{array}{l} X = \theta + E, \\ E \sim D(0, \sigma^2) \end{array}\right\} \tag{11.1}$$

and

$$X_k = \theta + E_k, \qquad k = 1, \ldots, n,$$
$$E_k \sim D(0, \sigma_k^2).$$

We look for a test statistic that has a known distribution and which will be sensitive to high or low values of X. Consider the statistic

$$X - \sum_k a_k X_k,$$

where $\sum_k$ indicates $\sum_{k=1}^n$. This variate has mean $\theta - \sum_k a_k \theta$, so to eliminate dependence of the distribution on θ we must set $\sum_k a_k = 1$. To maximize sensitivity to values of X that are high or low, we then choose to minimize the variance of $\sum_k a_k X_k$, which is achieved by setting a_k to be inversely proportional to the variance of X_k, (see, e.g., Hodges and Lehmann (1964), Secs. 10.1, 10.4). So the ideal test statistic is

$$X - \frac{\sum_k X_k/\sigma_k^2}{\sum_k 1/\sigma_k^2}.$$

This variate has mean 0 and, under the hypothesis H, has variance

$$\hat{\sigma}^2 + \frac{1}{\sum_k 1/\sigma_k^2}.$$

Also, to a first approximation it has a normal distribution. Therefore, if H is true, the variate

$$\frac{X - \frac{\sum_k X_k/\sigma_k^2}{\sum_k 1/\sigma_k^2}}{\sqrt{\hat{\sigma}^2 + \frac{1}{\sum_k 1/\sigma_k^2}}}$$

11.1 Examining the uncertainty statement

approximately has the standard normal distribution. Similarly, if H is true, the observable variate

$$Z \equiv \frac{X - \frac{\sum_k X_k/\hat\sigma_k^2}{\sum_k 1/\hat\sigma_k^2}}{\sqrt{\hat\sigma^2 + \frac{1}{\sum_k 1/\hat\sigma_k^2}}}$$

also approximately has the standard normal distribution N(0, 1). Here each unknown σ_k^2 has been replaced by its estimate $\hat\sigma_k^2$.

The central 2.5% of the standard normal distribution lies in the interval $[-0.0313, 0.0313]$ and the most extreme 2.5% of the standard normal distribution lies outside the interval $[-2.24, 2.24]$ so, if we choose a significance level of 0.05, the hypothesis H is to be rejected if the realization of Z, which is

$$z \equiv \frac{x - \frac{\sum_k x_k/\hat\sigma_k^2}{\sum_k 1/\hat\sigma_k^2}}{\sqrt{\hat\sigma^2 + \frac{1}{\sum_k 1/\hat\sigma_k^2}}}, \tag{11.2}$$

has magnitude less that 0.0313 or greater than 2.24. This hypothesis stands for the hypothesis that $\sigma_B^2 = \hat\sigma_B^2$ because we have accepted the accuracy of $\hat\sigma_A^2$. Thus, if $|z| \le 0.0313$ we conclude that $\sigma_B^2 < \hat\sigma_B^2$, while if $|z| > 2.24$ we conclude that $\sigma_B^2 > \hat\sigma_B^2$.

If our interest is only in the possibility that we have *underestimated* σ_B^2 then only high magnitudes of the test statistic are relevant, in which case (when the significance level is 0.05) we conclude that $\sigma_B^2 > \hat\sigma_B^2$ if $|z| > 1.96$.

Adjustment for application with many laboratories

The measurement comparison will be carried out to examine the uncertainty statements of *all* the participating laboratories, not just ours. The corresponding tests will usually focus on whether the uncertainty statements are too small. The principles of simultaneous inference discussed in Chapter 10 become relevant. If the uncertainty statements of all $n + 1$ laboratories are to be assessed in this way then, by the Bonferroni inequality, the significance level can be set to $0.05/(n + 1)$. The conclusion that any single estimate $\hat\sigma_k^2$ is too small will then have the same weight as the same conclusion would have in a stand-alone test of '$\sigma^2 = \hat\sigma_k^2$' with a significance level of 0.05.

Goodness-of-fit tests

The procedure above tests the equality of the estimate $\hat{\sigma}_B^2$ with the actual variance σ_B^2. The hypothesis tested '$\sigma_B^2 = \hat{\sigma}_B^2$' is known as a 'simple hypothesis' because it involves a single hypothesized value of a parameter. However, the parameter σ_B^2 exists on a continuous scale, so it is impossible for it to be exactly equal to any specified number, such as $\hat{\sigma}_B^2$. Thus this hypothesis is necessarily false, and the correct answer to the test is already known! So what is the point of carrying out the test? In our situation, if $|z| < 0.0313$ we conclude that $\sigma_B^2 < \hat{\sigma}_B^2$, in which case we have, in effect, tested and rejected the hypothesis that $\sigma_B^2 \geq \hat{\sigma}_B^2$ at a significance level of 0.025, this probability applying when $\sigma_B^2 = \hat{\sigma}_B^2$. Similarly, if $|z| > 2.24$ we conclude that $\sigma_B^2 > \hat{\sigma}_B^2$, in which case we have rejected the hypothesis that $\sigma_B^2 \leq \hat{\sigma}_B^2$ with significance level 0.025. In effect, we are simultaneously conducting tests of the hypotheses '$\sigma_B^2 \leq \hat{\sigma}_B^2$' and '$\sigma_B^2 \geq \hat{\sigma}_B^2$', neither of which is a simple hypothesis and so neither of which we know in advance to be false. Thus there is some sense in the procedure.

But even so, what is the value of concluding that $\sigma_B^2 \geq \hat{\sigma}_B^2$, say? Our goal is to determine whether $\hat{\sigma}_B^2$ is sufficiently close to σ_B^2 for our measurement technique to be fit for purpose, and the tests that we have described do not answer this question. One cause is that we have not specified what we mean by 'fit for purpose' and another is that we have tested the exactness of a specified relationship, not the degree of approximation in that relationship.

These comments are particularly relevant to so-called 'goodness-of-fit' tests, where statistical models are put forward as descriptions of the mechanisms by which data arise and then actual data are used to test whether these descriptions are accurate. The logic behind such tests is curious. While some models are undoubtedly valuable, the claim that a statistical model exactly describes a data-generating process is fanciful. This is reflected in a familiar maxim in statistics, 'all models are wrong, but some are useful' (e.g. Box and Draper (1987), p. 424), from which we can state that 'some models are useful, but all are wrong'. Despite the truth of this maxim, the hypothesis formally tested in a goodness-of-fit test is the hypothesis that the model is exact! Instead of examining the quality of fit of the model, which is what the phrase 'goodness of fit' will mean to many uninitiated readers, the test actually assesses the 'possibility of exactness' of the model. And if we accept that 'all models are wrong' then we already know the answer to the question that the test addresses!

The practice of goodness-of-fit testing seems based on the idea that if there is insufficient evidence in the data to show that the model is wrong, then the model cannot be very wrong. While this may be a useful idea, it is not logical. On one hand, an inadequate model might pass the test because the amount of data we possess is too small for the test to have high probability of resulting in rejection, i.e. high power. On the other hand, a model that is perfectly adequate for our purpose could fail a goodness-of-fit test very badly just because the data are sufficiently numerous to provide incontrovertible evidence that the model is inexact. So the test is not a direct assessment of the quality of a model, and the phrase 'goodness-of-fit' seems to be a misnomer.

This criticism of tests of simple hypotheses, and especially tests of goodness-of-fit, is one reason why few hypothesis tests are described in this book. Tests with a stronger logical basis seem required. In particular, for a test to be meaningful in logic, the hypothesis tested has to be potentially true. In our case, the hypothesis of interest, '$\sigma_B^2 = \hat{\sigma}_B^2$', cannot possibly be true because both σ_B^2 and $\hat{\sigma}_B^2$ exist on a continuous scale. A better test for assessing whether $\hat{\sigma}_B^2$ was an adequate approximation to σ_B^2 would be a test of the hypothesis '$|\sigma_B^2 - \hat{\sigma}_B^2| \leq \delta$', where δ was some suitable figure related to our purpose. In effect, this test is conducted by finding if δ or $-\delta$ is contained between the limits of a realized confidence interval for σ_B^2.

Confidence interval for σ_B^2

Instead of testing σ_B^2, we might prefer to estimate it using a confidence interval, which will be found from a confidence interval for σ^2. A 95% confidence interval for σ^2 is formed by including all values of σ^2 for which the above test of H : $\sigma^2 = \hat{\sigma}^2$ would be passed, i.e. by inverting the hypothesis test. So an approximate 95% confidence interval for σ^2 is formed from all candidate values σ_c^2 satisfying

$$0.0313 < \left| \frac{X - \frac{\sum_k X_k/\hat{\sigma}_k^2}{\sum_k 1/\hat{\sigma}_k^2}}{\sqrt{\sigma_c^2 + \frac{1}{\sum_k 1/\hat{\sigma}_k^2}}} \right| < 2.24.$$

Therefore, a realized approximate 95% confidence interval for σ^2 is made up of all values of σ_c^2 for which

$$\frac{d^2}{2.24^2} < \sigma_c^2 + \frac{1}{\sum_k 1/\hat{\sigma}_k^2} < \frac{d^2}{0.0313^2}$$

with

$$d \equiv x - \frac{\sum_k x_k/\hat{\sigma}_k^2}{\sum_k 1/\hat{\sigma}_k^2}.$$

The limits of a corresponding interval for σ_B^2 are found by subtracting $\hat{\sigma}_A^2$ from the limits of this interval. So a realized approximate 95% confidence interval for σ_B^2 is

$$\left[0.199\,d^2 - \frac{1}{\sum_k 1/\hat{\sigma}_k^2} - \hat{\sigma}_A^2,\ 1021.7\,d^2 - \frac{1}{\sum_k 1/\hat{\sigma}_k^2} - \hat{\sigma}_A^2\right]. \tag{11.3}$$

The limits of this interval differ by at least four orders of magnitude! Therefore, unless the upper limit happens to be small, the comparison cannot be said to have been a useful means of estimating σ_B^2.

So the comparison is not a tool for independently demonstrating the accuracy of our assessment of the total variance associated with Type B errors. In the end, only trust in our identification and assessment of the Type B error components can convince another party of the sufficient accuracy of $\hat{\sigma}_B^2$. However, as we shall see in Section 11.2, the comparison does provide direct evidence of the value of e_B, which is the actual Type B error.

(If d^2 is small then one or both of the limits of (11.3) might be negative. Of course, this does not imply that σ_B^2 might be negative. Physical considerations mean that we can raise any negative limit to zero without reducing the success rate of the procedure, i.e. without invalidating the confidence-interval procedure. If both limits become zero then it is clear that we have encountered one of the 5% of occasions on which we accepted we would fail to bound σ_B^2. This is one situation where our postdata level of assurance in a realized interval differs from our predata level of confidence in the corresponding random interval. See Chapter 15 for detailed discussion of such a problem.)

11.2 Estimation of the Type B error

Our attention turns from the parent variance of the Type B error to the actual value of this error e_B. There are two different statistical problems that might be undertaken at this point. The first is the estimation of e_B without use of the figure $\hat{\sigma}_B^2$, one goal being to confirm that e_B and $\hat{\sigma}_B^2$ are consistent. The second possibility is the estimation of e_B after the accuracy of $\hat{\sigma}_B^2$ has been accepted. We shall discuss only the first problem, where e_B is estimated without use of the figure $\hat{\sigma}_B^2$. The estimate of e_B obtained may then be compared with $\hat{\sigma}_B^2$ to support the accuracy of $\hat{\sigma}_B^2$ as an estimate of σ_B^2.

The error, e_B, is the outcome of a variate E_B, so the probability statements that are relevant in the estimation of e_B by a confidence interval are those that are applicable conditional on the event that E_B takes the value e_B. This error e_B is then seen as a fixed input to the analysis, and we will rename it b for 'bias'. Thus

$$b \equiv e_B.$$

The Type B error variance σ_B^2 influences the generation of b but plays no part in probability statements applicable after the event $E_B = b$, which are the probability statements involved in the conditional confidence interval.

11.2 Estimation of the Type B error

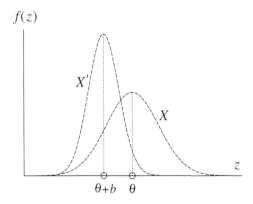

Figure 11.1 The distributions of the unconditional estimator X and the conditional estimator X' for a negative value of b.

Let X' denote the variate X on the condition that $E_B = b$. The model (11.1) becomes

$$X' = \theta + b + E_A,$$
$$E_A \sim D(0, \sigma_A^2),$$

with $\hat{\sigma}_A^2$ being our estimate of σ_A^2. We wish to develop a suitable estimator of b using this model and the model

$$X_k = \theta + E_k, \qquad k = 1, \ldots, n,$$
$$E_k \sim D(0, \sigma_k^2)$$

for which $\hat{\sigma}_k^2$ is the estimate of σ_k^2. Figure 11.1 gives an example of the distribution of X and depicts a corresponding distribution of X' for a negative value of b. The variance of X' is less than the variance of X by σ_B^2.

Let $\hat{b}$ denote the estimate that we shall obtain of b, and let $\check{b}$ denote the corresponding estimator. We consider estimators of the form

$$\check{b} = cX' + \sum_k c_k X_k, \tag{11.4}$$

which are *linear estimators* because the observable variates enter linearly. The only observable variate with a distribution involving b is X', so if b is to be fully represented in (11.4) then the coefficient c must be unity. Subsequently, if θ is to be absent from (11.4) the c_k coefficients must satisfy $\sum_k c_k = -1$. When these requirements are met the estimator $\check{b}$ is unbiased. The variance of $\check{b}$ is then minimized when $-c_k$ is set inversely proportional to the variance of X_k, which is σ_k^2. Therefore, the ideal estimator of b would be

$$X' - \frac{\sum_k X_k/\sigma_k^2}{\sum_k 1/\sigma_k^2}. \tag{11.5}$$

The variance σ_k^2 is unknown and is replaced by its estimate, which is treated as a constant. So the preferred estimator is

$$\check{b} = X' - \frac{\sum_k X_k/\hat{\sigma}_k^2}{\sum_k 1/\hat{\sigma}_k^2}$$

and the corresponding estimate is

$$\hat{b} = x - \frac{\sum_k x_k/\hat{\sigma}_k^2}{\sum_k 1/\hat{\sigma}_k^2} \tag{11.6}$$

(which is the same as the quantity d in Section 11.1).

The variate in (11.5) is unbiased, has variance $\sigma_A^2 + (\sum_k 1/\sigma_k^2)^{-1}$ and, to a first approximation, has a normal distribution. Therefore, a realized approximate conditional 95% confidence interval for b is

$$\left[\hat{b} - 1.96\,\hat{\sigma}_p(\hat{b}),\ \hat{b} + 1.96\,\hat{\sigma}_p(\hat{b})\right],$$

where

$$\hat{\sigma}_p(\hat{b}) = \sqrt{\hat{\sigma}_A^2 + \frac{1}{\sum_k 1/\hat{\sigma}_k^2}}. \tag{11.7}$$

This is a conditional confidence interval because it holds on the condition that $E_B = b$.

Test of consistency

To obtain a test of the hypothesis '$\sigma_B^2 = \hat{\sigma}_B^2$' through this approach, we refer the estimate $\hat{b}$ to the parent distribution that it would have if this hypothesis were true. On the condition that $E_B = b$, this distribution is approximately normal with mean b and variance $\hat{\sigma}_p^2(\hat{b})$. Also, if this (null) hypothesis were true then the parent distribution of b itself would have mean 0 and variance $\hat{\sigma}_B^2$. Thus the unconditional null parent distribution of $\hat{b}$ is approximately normal with mean 0 and standard deviation

$$\sqrt{\hat{\sigma}_B^2 + \hat{\sigma}_A^2 + \frac{1}{\sum_k 1/\hat{\sigma}_k^2}} = \sqrt{\hat{\sigma}^2 + \frac{1}{\sum_k 1/\hat{\sigma}_k^2}}.$$

This quantity is the denominator in (11.2) and the estimate $\hat{b}$ is the numerator. So the resulting test of consistency is exactly the same as that described in Section 11.1.

11.3 Experimental bias

In many situations each component of the Type A error will be random and each component of the Type B error will be fixed. In such a situation, b is the *experimental bias* and the estimation of b takes on greater significance because of the possibility of applying a useful correction in future measurements. In effect, we will have used the inter-laboratory comparison to calibrate our measurement technique. In this section, we consider the possibility of making a correction for this bias. This correction will be applicable in future measurements made using the same technique.

As explained in Chapter 4, if fixed errors are to be treated probabilistically then (i) the process of measurement must be seen to include background steps and (ii) the concept of repeated experimentation must be broadened to one of repeated measurement. Under this idea that fixed errors can be 'randomized', the variate X has mean θ, which means that – according to the statistician's definition of bias – the measurement is unbiased. But during the construction of the measurement system the variate E_B took the unknown value b, which becomes a bias conditional on the event '$E_B = b$'. So, to a statistician asked to accept the idea that fixed errors can be treated probabilistically, b is not strictly a bias but is a *conditional* bias.

Correcting for the experimental bias

Let X_f, θ_f and $E_{A,f}$ now represent the estimator, target value and random-error variate in a *future* measurement of the same type. We can write

$$X_f = \theta_f + b + E_{A,f},$$

where b is the bias that was examined in the comparison. The distribution of $E_{A,f}$ will be the same as the distribution of the random-error variate applicable in the comparison. So the model conditional on the existence of that bias is

$$x_f \leftarrow N(\theta_f + b, \sigma_A^2),$$

where x_f is the realization of X_f. The best estimate of b is the quantity $\hat{b}$ given by (11.6), and an approximation to its parent variance is the quantity $\hat{\sigma}_p(\hat{b})^2$ given in (11.7). The figure $\hat{b}$ was drawn from an approximately normal distribution with mean b and variance $\hat{\sigma}_p(\hat{b})^2$, so

$$x_f - \hat{b} \xleftarrow{\text{approx}} N\left(\theta_f, \sigma_A^2 + \hat{\sigma}_p(\hat{b})^2\right).$$

Thus $x_f - \hat{b}$ is an approximately unbiased estimate of θ_f, and $\hat{\sigma}_p(\hat{b})^2$ is a figure that should be seen as the new Type B error variance. The component of parent variance due to the fixed error e_B has been replaced by our approximation to the parent variance of the estimate of bias, which will almost always be smaller. The application of a correction has allowed us to reduce the amount of uncertainty to be quoted with future measurement estimates.

In practice, we would be unwilling to make such a correction without fully understanding our measurement technique, and we would be unlikely to claim this full understanding unless we trusted the figure $\hat{\sigma}_B^2$. So, in practice, the statement $b \leftarrow D(0, \hat{\sigma}_B^2)$ would be incorporated into the model when forming an estimate of bias that will be used as a correction. Using this additional information must lead to an estimate of b with even smaller uncertainty than $\hat{b}$. As implied earlier, we do not consider this alternative estimation problem in this book (see Willink (2012b)).

Estimating differences between biases

There are some situations where we might wish to estimate the difference between the biases of two different techniques or laboratories. Let b_1 and b_2 be the biases, let X_1 and X_2 be the variates for the measurement results when these procedures are used to measure the same target value θ, and let x_1 and x_2 be the numerical results. The relevant model consists of the relationships

$$X_1' = \theta + b_1 + E_{1A},$$
$$E_{1A} \sim D(0, \sigma_{1A}^2)$$

and

$$X_2' = \theta + b_2 + E_{2A},$$
$$E_{2A} \sim D(0, \sigma_{2A}^2),$$

where X_1' is the variate X_1 on the condition that $E_{1B} = b_1$ and X_2' is the variate X_2 on the condition that $E_{2B} = b_2$. The realizations of X_1' and X_2' are x_1 and x_2.

Although results from other techniques or laboratories might be available, there is no better estimator of $b_1 - b_2$ than $X_1' - X_2'$, which conditionally is unbiased and has variance $\sigma_{1A}^2 + \sigma_{2A}^2$. No other linear combination of the X_k' variates will be conditionally unbiased and have smaller variance. This variance will be estimated by $\hat{\sigma}_{1A}^2 + \hat{\sigma}_{2A}^2$. So $x_1 - x_2$ is the preferred estimate of $b_1 - b_2$ and the interval

$$\left[x_1 - x_2 - 1.96\sqrt{\hat{\sigma}_{1A}^2 + \hat{\sigma}_{2A}^2},\ x_1 - x_2 + 1.96\sqrt{\hat{\sigma}_{1A}^2 + \hat{\sigma}_{2A}^2} \right]$$

is a realized approximate 95% confidence interval for $b_1 - b_2$.

11.4 Complicating factors

The methods described here have been presented without questioning either the accuracy of our estimate of σ_A^2 or the accuracy of the estimate of the total variance σ_k^2 for another participant in the comparison. Incorporating the idea that these estimates will themselves be inexact would complicate the analysis considerably. Whether such complication is required depends on the actual purpose of the comparison and how the results are to be used. Unless there is a predetermined course of action to be taken if the test of consistency is failed then the procedure, with its arbitrary significance level, lacks practical meaning. As a result, there seems to be no need for a more complicated procedure. Also, the comparison is relatively poor as a means of estimating σ_B^2, and this raises the question of whether complication would be justified in practice.

As we have seen, the comparison is more useful as a means of estimating e_B, especially when this error is interpretable as the total fixed error. However, if no correction is to be made for a fixed error in subsequent measurements then there seems little point in having a formal estimate of it, and the simple analysis given seems sufficient. An unwillingness to make such a correction in practice would seem to have only one legitimate source, which is a lack of confidence in the idea that e_B is actually constant. But it is worth noting that, in the international metrology community, the means of linking a regional measurement comparison to a global measurement comparison might well involve the assumption that the bias or 'degree of equivalence' of a laboratory taking part in both comparisons is constant in the intervening period (e.g. Elster *et al.* (2010)).

In summary, we may say that a measurement comparison can provide direct evidence about the value of the total Type B error but can only provide corroborating evidence about the parent variance of this error. Further complication of the analysis does not seem justified unless this evidence is to be acted upon, which could involve the application of a correction in future measurements. This correction would usually result in a smaller uncertainty of measurement.

12
Other philosophies

In this book, the frequentist concept of estimation of a target value by a confidence interval has been reinterpreted and extrapolated to accommodate the realities of measurement. I like to think that the resulting methodology follows logically from the principles put forward in Part I, which were summarized in Section 6.1. Not everyone will agree with the approach taken, but to disagree seems to require finding fault with one or other of those principles.

This chapter outlines other points of view about measurement. It contains a brief discussion of the use of worst-case values for systematic errors, a short introduction to the concept of 'fiducial' inference and a description of different types of Bayesian statistics. It closes with an examination of the implications of denying the existence of a target value.

12.1 Worst-case errors

One traditional approach to the statement of accuracy in measurement involves the imputation of worst-case values for systematic errors (Colclough, 1987; Grabe, 2005). This point of view accords with the strict frequentist principle that the advertised success rate is to be reached whatever the values of all non-random influences in the measurement situation. Adherence to this principle ensures that the overall success rate in any set of problems also reaches the advertised value. Although I have argued that we can relax this view to envisage systematic errors being drawn from distributions, I have a lot of respect for this idea of using worst-case values for systematic effects. If this approach was popular then I would be someone seeking to support it. However, it is not popular, and the idea that systematic errors can be treated probabilistically is at the heart of international guidelines (Kaarls, 1980; BIPM *et al.*, 1995). Undoubtedly, one reason for this lack of popularity is the fact that the interval obtained using worst-case values is often long compared to that obtained with a probabilistic treatment: the interval is regarded as being unnecessarily conservative or unrealistic. However, such a criticism can itself be criticized: what does 'unrealistic' actually mean and what level of performance would correspond to a 'necessary' degree of conservatism? This brings us back to the central question: *What is to be understood when a claim of 95% assurance is attached to an interval of measurement uncertainty?*

12.2 Fiducial inference

The concept of 'fiducial probability' was put forward by Sir Ronald Fisher during an important period in the history of mathematical statistics (Fisher, 1930; Edwards, 1976, 1983). The presentation of the fiducial idea led indirectly to the development of the theory of confidence intervals by Jerzy Neyman (Neyman, 1937), which is widely accepted.

The fiducial approach involves the construction of a probability distribution to describe the value of an unknown but fixed quantity of interest, i.e. the relevant parameter of the problem. The probability distribution can be formed solely from the data and the model of the data-generating mechanism: no 'prior distribution' for the parameter is used. Thus, Fisher was considering a way of forming a probability distribution for a fixed unknown quantity without involving the controversial idea of a 'prior distribution'.

The basic implication of fiducial probability seems best illustrated by example. Suppose X is normally distributed with mean θ and known variance σ^2 and suppose x is the realization of X. Then, according to the fiducial idea, after observing x we can consider θ to have the normal distribution with mean x and variance σ^2. This would be called the 'fiducial distribution' of θ. The symmetry of the normal distribution hides the fact that a 'reflection' of the error distribution is involved: the sampling distribution of the error $X - \theta$ is reflected about the origin and becomes the fiducial distribution of the quantity $\theta - x$. So the subject of the probability statement is changed from the estimator X to the unknown parameter θ.

The presence of such a reflection is seen in a more general description. Consider a measurement with target value θ. If the measurement-error variate $E \equiv X - \theta$ has probability density function $f_E(\cdot)$ and if x is the measurement result then the fiducial density function of θ is

$$f_\theta(z) = f_E(x - z). \tag{12.1}$$

Figure 12.1 shows this density function when a quantity θ is measured using a method known to incur positive error drawn from an exponential distribution with parameter λ (as in Figure 4.7). Here, the exponential curve is increasing rather than decreasing because the sign of z on the right-hand side of (12.1) is negative.

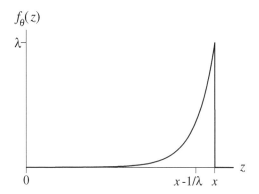

Figure 12.1 The fiducial distribution of θ after observing x when the error is drawn from the exponential distribution with parameter λ shown in Figure 4.7.

So, if applying the fiducial approach in a problem where there is a single unknown parameter θ, a probability distribution for θ is formed from an estimate x and from the sampling distribution of the error in the corresponding estimator X. This simplicity is appealing, and the fiducial way of thinking may be unwittingly adopted by many scientists when they choose to make probability statements about a parameter after observing the experimental data. However, the apparent intuitiveness of the method hides the facts that difficulties exist with generalizing the fiducial idea[1] and that it is not accepted in the statistical community. In fact, as long ago as 1978 the fiducial argument was described in a review article as 'essentially dead' (Pedersen, 1978). To construct a probability distribution for a parameter using principles that many see as proper requires a Bayesian approach, as is described in Section 12.3.

Despite this unpopularity, a number of papers involving fiducial probability have appeared in the metrology literature (Wang and Iyer, 2005, 2006, 2008, 2009, 2010; Hannig et al., 2007; Frenkel, 2009). In those papers, the authors tend to evaluate their methods using frequentist principles, where success rates are examined at fixed values of the unknown parameters. In this way the fiducial approach is being treated as a tool for achieving an appropriate success rate rather than as a philosophy of inference. If a tool enables the goal to be achieved then its use is justified but, arguably the range of problems for which a fiducial approach performs well and is practical seems too limited for the theory to be the basis of a system of analysis. Indeed, Fisher only presented the fiducial approach for use in a subset of problems (Edwards, 1976). Further discussion on fiducial inference in metrology (along with material on frequentist and Bayesian inference) is given by Guthrie et al. (2009).

---

[1] For example applying the rules of probability to the fiducial distribution for the mean μ of a normal distribution does not give the fiducial distribution that would be obtained directly for μ^2 (Pedersen, 1978).

So fiducial inference involves the concept of attributing a probability distribution to an unknown but fixed quantity. This might be called the *distributed-measurand* concept (Willink, 2010a). This concept is also foundational in the field of Bayesian statistics, which is now discussed.

12.3 Bayesian inference

The principal competitors to the frequentist point of view in statistics are the various points of view found in *Bayesian statistics*. The essential idea of Bayesian statistics is that a probability is attributed to every relevant event whose outcome is unknown, whether that event has already taken place or has yet to take place. This probability describes someone's strength of belief about the event or hypothesis, and it is not necessary that the event be an outcome in a repeatable experiment for the assignment of this probability to be deemed meaningful.

So the Bayesian approach permits direct statements of probability to be made about all relevant unknown quantities, such as the achievements of Mallory and Irvine in (3.1), Newton's constant G in (3.2) and the target value θ in a general measurement. This might be seen as an advantage, but we must recognize that adopting this approach *forces* such statements to be made both before and after obtaining experimental data, and these statements must constitute complete probability distributions. For example before conducting any experiment to estimate any target value θ, a probability distribution describing belief about θ must be specified. Likewise, before conducting any procedure that will be repeated to obtain a set of results for averaging, a probability distribution is required to describe belief about the unknown variance in the output of the procedure.

Thus (3.1) and (3.2) are statements that would *potentially* be deemed meaningful by a Bayesian statistician. I write 'potentially' because there is division amongst Bayesian statisticians. Indeed, it is difficult to give any single definition of Bayesian inference other than to emphasize the ideas that in Bayesian statistics (i) unknown parameters are regarded as being variates (random variables) and (ii) systematic use is made of Bayes' theorem (Ledermann and Lloyd, 1984; Marriott, 1990).

In this section, we discuss the basic principles of *subjective Bayesian statistics* and very briefly introduce the field of *objective Bayesian statistics*, which is described more fully in Chapter 13. We also discuss the idea of a frequency-based Bayesian analysis, which might have a legitimate role to play. First, we describe in broad terms the mechanics of Bayesian estimation of a fixed quantity.

The elements of a Bayesian analysis

There are two basic ingredients in the estimation of an unknown fixed quantity, θ, using a Bayesian method. The first is a probability density function, $f_{\text{prior},\theta}(z)$, that describes belief about the value of θ before the experiment is commenced. This constitutes the *prior distribution* of θ. The second is a probability density function, $f(\text{random data} | \theta = z)$, that describes belief about the possible data to be obtained if θ were to have the dummy value z. Actual data will be obtained in the experiment and this function will then become $f(\text{actual data} | \theta = z)$, which is a function of z alone and which is called the *likelihood function*. These ingredients are then combined using Bayes' theorem, which can be written as

$$f_{\text{post},\theta}(z) \propto f_{\text{prior},\theta}(z) f(\text{actual data} | \theta = z), \tag{12.2}$$

to give a *posterior distribution* with density function $f_{\text{post},\theta}(z)$ that, according to theory, describes the belief about the value of θ after the experiment. This posterior density function is the fundamental output of the Bayesian procedure, and it is said to contain all that is believed about θ after the experiment. Any interval containing 95% of this distribution is called a *95% credible interval* for θ, and it is regarded as 95% probable that θ lies in any such interval. The posterior distribution would then act as the prior distribution of θ for the next experiment.

Example 12.1 A certain course of action can be taken if it is concluded that there is at least 95% probability that a fixed quantity θ exceeds the level 1000. The probability distribution attributed to θ is the normal distribution with mean 1200 and standard deviation 200, so the initial probability that θ exceeds 1000 is deemed to be 84%. More information can be gained by measuring θ, and it may be that after measurement the calculated probability that θ exceeds 1000 will be greater than 95%, in which case the action can then be taken. The measurement procedure is known to incur a normally distributed error with mean 0 and standard deviation 150. A measurement is made, and the result is $x = 1250$. Can the course of action be taken?

The prior probability distribution of θ is $N(1200, 200^2)$. So the prior density function of θ is

$$f_{\text{prior},\theta}(z) \propto \exp\left\{-\frac{1}{2}\left(\frac{z-1200}{200}\right)^2\right\} \tag{12.3}$$

(and the corresponding probability that θ exceeds 1000 is $\Pr_{\text{prior}}(\theta > 1000) = 0.84$). The likelihood function is

$$f(1250 | \theta = z) \propto \exp\left\{-\frac{1}{2}\left(\frac{1250-z}{150}\right)^2\right\}.$$

Application of Bayes' theorem (12.2) shows that the posterior density function satisfies

$$f_{\text{post},\theta}(z) \propto \exp\left\{-\frac{1}{2}\left(\frac{z-1232}{120}\right)^2\right\}, \tag{12.4}$$

12.3 Bayesian inference

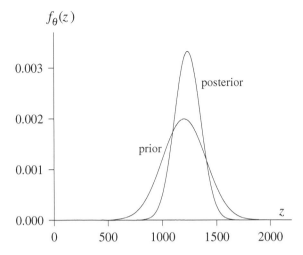

Figure 12.2 The prior and posterior distributions of θ.

which implies that the probability distribution attributed to θ after observing the measurement result is $N(1232, 120^2)$. Therefore the posterior probability that θ exceeds 100 is $\Pr_{\text{post}}(\theta > 1000) = 0.97$. So the measurement has provided the extra information required for the course of action to be taken.

Figure 12.2 shows the prior and posterior distributions corresponding to (12.3) and (12.4). In accordance with the idea that making the measurement provides information, the variance of the distribution attributed to θ has been reduced.

As a mathematical relationship, Bayes' theorem is accepted by all statisticians: the subject of dispute is the meaningfulness of the prior distribution. The frequentist statistician is unlikely to accept a prior distribution for a parameter as being meaningful unless this distribution describes a known frequency distribution from which the parameter has arisen. This reluctance is indicative of the main difference between Bayesian statistics and frequentist statistics, which – as was argued in Section 3.1 – is the universe of the events thought permissible to be attributed probabilities. This difference relates to the breadth of the idea that probability is 'degree of belief'.

Probability as degree of belief

The idea that a numerical probability is interpreted as a degree of belief is at the heart of Bayesian statistics, but this idea is of little value if 'degree of belief' itself remains undefined. As suggested in Section 3.1, standard definitions of degree of belief are based around the idea of betting for 'expected gain'. However, a mathematical expectation is only a relevant quantity when a long-run view is taken. So, although many Bayesian statisticians will claim that their concept of probability allows them to associate a probability with an unrepeatable one-of-a-kind event, that event must be seen as part of some hypothetical series of events within the person's experience. For example attributing a probability of 95%, or 19 out of 20, to some event makes no sense if the person does not have the concept that life involves at least 20 equally reliable opportunities or equally plausible ideas. (Example 4.3 seems relevant at this point.)

The definition of personal probability given in Definition 3.1, which also acts as a definition of 'degree of belief', implies that a certain long-run behaviour is anticipated. A Bayesian approach would be more appealing if this long-run meaning of degree of belief was emphasized. A helpful statement was made by M. J. Bayarri and the well-known Bayesian author J. O. Berger in a prestigious statistics journal (Bayarri and Berger, 2004). They wrote:

If (say) a Bayesian were to repeatedly construct purported 90% credible intervals in his or her practical work, yet they only contained the unknowns about 70% of the time, something would be seriously wrong.

This statement acknowledges the accountability of Bayesian methods to the requirement of an acceptable long-run success rate. The truth of this statement will be self-evident to many readers. (If my experience with a referee is anything to go by, some readers will misinterpret this statement of Bayarri and Berger. They will think that it is only meaningful if the success rate can be examined, which requires knowing the unknowable target values. But the statement holds whether we know the success rate or not. Something can be wrong without us knowing about it!)

The idea in Bayesian statistics that probability is degree of belief is often said to provide a contrast with frequentist statistics. However, as was pointed out in Section 3.1, the frequentist view of probability also involves degree of belief. A probability featuring in a frequentist model is interpretable as a strength of belief about the next outcome of a repeatable procedure: the degree of belief is simply the anticipated frequency under repetition. Instead of focusing on the idea of belief as being peculiar to Bayesian statistics, it is more helpful to see the difference between Bayesian and frequentist statistics as relating to the *objects* of this belief. The Bayesian statistician allows, and requires, probability statements to be made about the unknown outcomes of all relevant events, whether these occur in the past, present or future.

Subjective Bayesian statistics

We have emphasized that Bayesian statisticians see probability as degree of belief. Those who have a subjectivist view of probability take this idea seriously.

12.3 Bayesian inference

Quite correctly in my view, they see belief as being something that is personal. Consequently, the beliefs represented by both the prior distribution and the likelihood model in a subjective Bayesian analysis are to be the actual beliefs (before acquisition of the data) of the individual who is going to use the results of the analysis.

These beliefs can be evaluated in a thought experiment. If someone is interested in estimating θ then that person might be asked, for a specific number z, 'what is the maximum amount you would pay to receive 1 unit if θ is less than or equal to z?' The person's answer, according to Definition 3.1, is the probability that he or she attributes to the hypothesis '$\theta \leq z$'. Answering this question for many values of z before acquisition of any experimental data enables the construction of the prior probability distribution function $F_{\text{prior},\theta}(z)$ that the person attributes to θ. Differentiating this distribution function gives the prior probability density function $f_{\text{prior},\theta}(z)$. By such questioning, the statistician can elicit an appropriate probability distribution to represent the person's prior belief.

The following example involves a converse kind of question.

Example 12.1 (revisited, with a subjective prior distribution) The prior distribution attributed to θ by the person making the decision is $N(1200, 200^2)$. The person constructed this distribution by asking himself–herself the converse question 'what is the smallest value z for which I would pay up to $p = 0.1$ unit to receive 1 unit if $\theta \leq z$?' and then asking this question with the values $p = 0.3, 0.5, 0.7, 0.9$ also. The five answers are $z = 950, 1100, 1200, 1300, 1450$ respectively. So

$$F^{-1}(0.1) = 950,$$
$$F^{-1}(0.3) = 1100,$$
$$F^{-1}(0.5) = 1200,$$
$$F^{-1}(0.7) = 1300,$$
$$F^{-1}(0.9) = 1450,$$

where $F(z)$ is the probability distribution describing the person's actual belief. The person's prior belief about the value of θ therefore seems well approximated by the normal distribution $N(1200, 200^2)$, for which

$$F^{-1}_{\text{prior},\theta}(0.1) = 944,$$
$$F^{-1}_{\text{prior},\theta}(0.3) = 1095,$$
$$F^{-1}_{\text{prior},\theta}(0.5) = 1200,$$
$$F^{-1}_{\text{prior},\theta}(0.7) = 1305,$$
$$F^{-1}_{\text{prior},\theta}(0.9) = 1456.$$

For simplicity, this approximating distribution is taken as the prior distribution for θ. Figure 12.3 shows the five points from the function $F(z)$ and shows the prior

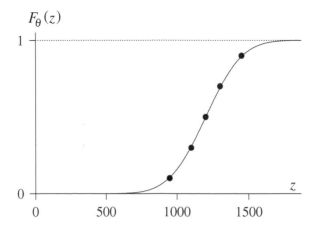

Figure 12.3 Five points on the person's actual prior 'distribution function' of belief about θ, $F(z)$, and the distribution function of $N(1200, 200^2)$, which is the prior distribution function $F_{\text{prior},\theta}(z)$ attributed to θ in the analysis.

distribution function $F_{\text{prior},\theta}(z)$ taken to represent the prior belief of the person about θ.

Provided that 'degree of belief' is suitably defined – as in Definition 3.1 – and provided that the consequences of the definition are explored, subjective Bayesian statistics seems to have intellectual integrity. This is reflected in the idea that subjective Bayesian statistics is 'coherent', which loosely means that a 'rational' person applying the methodology cannot obtain contradictory or illogical results. The main criticism of subjective Bayesian statistics in science would seem to be that the results lack authority and relevance away from the individual whose beliefs have been quantified. We may well ask 'what has personal belief got to do with the reporting of scientific results?'

Objective Bayesian statistics

The subjectivity of belief is the Achilles' heel of Bayesian statistics as described so far. This subjectivity is unacceptable to many scientists. For this reason, some turn to so-called 'objective' Bayesian statistics, where the prior distributions employed are designed to represent ignorance or minimal information, or are intended to be neutral in some way. The idea that probability describes genuine belief becomes more difficult to see, and the idea of quantifying belief seems to give way to the idea of quantifying 'information'. Consequently, we must ask whether the probabilities calculated in objective Bayesian analyses actually obey Definition 3.1.

12.3 Bayesian inference

The field of objective Bayesian statistics is much more complex and problematic than that of subjective Bayesian statistics. It is the subject of Chapter 13, where a more detailed criticism of this approach is given.

Frequency-based Bayesian statistics

Consider a problem in which a fixed parameter θ is to be estimated. Making a prior probability statement about θ seems meaningful if this problem is a typical member of a set of problems in which the values of the parameter analogous to θ are distributed in a known way. This distribution will act as the prior probability distribution of θ. The term 'frequency-based Bayesian analysis' is a useful description for an analysis that uses such a prior distribution. The frequentist statistician is unlikely to find fault in the subsequent use of Bayes' theorem, for the probability statements found in the prior distribution will possess long-run interpretations.

> **Example 12.1 (revisited, with the prior distribution as a frequency distribution)** The prior distribution attributed to θ by the person making the decision is $N(1200, 200^2)$. This distribution is chosen because the fixed values analogous to θ in previous problems of this type have been found to be distributed in this way. There is nothing to suggest that the problem at hand is atypical, so θ can be thought of as having been drawn randomly from this normal distribution.

In essence, the idea being described is that of regarding the value of an unknown parameter θ as having been drawn from some known distribution. We have already employed this idea in Chapter 4 when justifying the randomization of systematic errors and in Section 5.1 when extending the concept of an ordinary confidence interval to that of an average confidence interval. However, we must draw a distinction between the frequentist representation of this idea and the Bayesian representation. The frequentist sees θ as being the realization of a variate Θ, so probability statements are made about Θ, not θ. The Bayesian makes probability statements about θ itself, and so sees θ as a variate. So the two paradigms remain incompatible even if every prior distribution is of this type.

Summary

This author's views about Bayesian analysis may be summarized as follows. The probability statements that are the outputs of a typical subjective Bayesian analysis seem to lack general relevance, while many of the probability statements that are the outputs of typical objective Bayesian analyses seem to lack genuine meaning. Only when the prior distributions involved can be interpreted as frequency distributions will the outputs necessarily obey a frequency interpretation, which is required and implied by our definitions of probability, Definitions 3.1 and 3.2. Even then the Bayesian and frequentist paradigms remain incompatible because the Bayesian sees the corresponding parameter as a variate but the frequentist would see it as the outcome of a variate.

12.4 Measurement without target values

A distrust of the idea that the measurand has a 'true value' seems ingrained in the minds of many measurement scientists. Perhaps the idea of a 'target value' θ might also be found unacceptable. However, as was argued in Section 1.2, if there is no θ then the measurement problem is not one of *estimating* θ and the classical and Bayesian ideas of a confidence interval and credible interval for θ are inapplicable. In that case, for the concept of probability to remain relevant to some idea of measurement uncertainty, we must find a different question to answer. Moreover, if there is no target value then what event or fact are we uncertain about?

In this section, we consider further the premise that there is no target value in measurement and we explore some consequences of this premise on the role of probability and the meaning of measurement uncertainty. I hope that if the reader perceives the discussion to be taking on a flavour of absurdity then he or she will see this as a fault in the premise, not a fault in the ensuing argument.

Accuracy and uncertainty

As implied in Section 1.1, a measurement may be defined as *a process of experimentally obtaining a value (or more than one) that can reasonably be attributed to a quantity* (see VIM (2012), clause 2.1). Obviously, this value is to be the 'best', by some criterion. However, if there is no θ then there is no concept of an unknown error $x - \theta$ and there is no obvious concept of 'accuracy'. So what becomes the meaning of 'best value'? Perhaps, it means the value that is most in keeping with the results that others would obtain when measuring the same quantity. If so, the only idea of correctness becomes one of agreement with others. A critic might then say that the idea of consensus is taking on more importance than any objective reality of the measurand.

The word 'uncertainty' means doubt, so if 'measurement uncertainty' is not to be a misnomer then there must be a thing that we are doubtful about, and this thing must be amenable to quantitative description. It seems to me that, in the absence of a target value, the only quantitative thing unknown is the distribution of results that would be obtained if other measurements were made of the measurand. So, without a target value, the concept of measurement uncertainty seems related to describing the spread of the results that would be obtained if our measurement (including the

12.4 Measurement without target values

background steps) were repeated. It addresses the question of what result I might obtain when measuring the same measurand again with recalibrated equipment and regenerated reference values. The implication is that *the standard uncertainty of measurement is the standard deviation of the set of results that would be obtained in a very large number of measurements* (see Dietrich (1973), p. 8). In the current era, where systematic effects are attributed parent distributions, the results referred to would be hypothetical, for we cannot in practice repeat the measurement with different values of the systematic errors. If this is our concept of measurement uncertainty then the interval of uncertainty will be a 'probability interval' or a 'tolerance interval', as now described. (See Mandel (1964).)

Probability intervals and tolerance intervals

Our context is now one of predicting the outcome of a variate Y. Let us envisage an interval that contains 95% of the distribution of Y. An appropriate name for this interval is *95% probability interval*.[2]

Definition 12.1 A *95% probability interval for a variate Y* with probability density function $f(y)$ is any interval $[c, d]$ for which $\int_c^d f(y)\,dy = 0.95$.

Thus, if q_p is the p quantile of a continuous variate Y then any interval $[q_p, q_{p+0.95}]$ with $0 \leq p \leq 0.05$ is a 95% probability interval for Y.

Example 12.2 An extremely large number of independent measurements are made of a quantity. The results are normally distributed with mean 95.2 and standard deviation 0.3. The sample is so large that these figures can be taken to be the mean and standard deviation of the underlying distribution. So (with $p = 0.025$) a 95% probability interval for the result of a future independent measurement is

$$\left[95.2 + 0.3\,\Phi^{-1}(0.025),\ 95.2 + 0.3\,\Phi^{-1}(0.975)\right],$$

with $\Phi(z)$ being the distribution function of $N(0, 1)$. The quantities $\Phi^{-1}(0.025)$ and $\Phi^{-1}(0.975)$ are the familiar numbers -1.96 and $+1.96$. So there is probability 95% that the next measurement result will fall in the interval $[94.6, 95.8]$.

Recall that a 95% confidence interval is a random interval with probability 0.95 of covering the fixed quantity being estimated. In contrast, a 95% probability interval is a fixed interval with probability 0.95 of covering the value to be taken by a random variable (a variate). So the concepts of a confidence interval and a probability interval are quite different. However, sometimes in the measurement literature

---

[2] The term 'prediction interval' is appealing, but it has a different meaning. See p. 189.

the limits of a probability interval are called 'confidence limits' (e.g. Hughes and Hase (2010), p. 25). This is unhelpful because it subsequently becomes more difficult for the correct meaning of a confidence interval to be appreciated.

Often our knowledge of a distribution comes from a sample that is small or moderate in size, in which case we do not know any quantile of the distribution. In this situation we can only have a specified level of 'confidence' that a calculated interval contains at least 95% of the distribution. So we turn back to the concept of confidence, but now the object of study is the distribution itself, not a parameter like the mean of the distribution. We consider the concept of a *tolerance interval*, which is a random interval that has a specified probability of covering at least a certain fraction of the distribution.

Definition 12.2 A *95%-content tolerance interval for a variate Y with confidence coefficient 99%* is a random interval $[A, B]$ such that

$$\Pr\left(\int_A^B f(y)\,dy \geq 0.95\right) = 0.99.$$

The interval $[A, B]$ is random because one or both of A and B are variates, the usual situation being where these variates are both functions of random sample data.

Example 12.3 The potential results of measurements of a quantity are known to be independently drawn from a normal distribution, but the mean and variance of this distribution are unknown. A predetermined number, n, of measurements are to be made, and the variates for the measurement results are $X_1, \ldots, X_n$. The random sample mean is $\bar{X} = \sum_{i=1}^{n} X_i/n$ and the random sample variance is

$$S^2 = \frac{\sum_{i=1}^{n}(X_i^2 - \bar{X})^2}{n-1}.$$

Let $\chi^2_{0.01,n-1}$ indicate the first percentile of the chi-square distribution with $n-1$ degrees of freedom, and set

$$k = 1.96 \times \sqrt{\frac{(n^2-1)/n}{\chi^2_{0.01,n-1}}}.$$

The interval with limits $\bar{X} \pm kS$ has probability approximately 0.99 of covering 95% of the unknown normal distribution (Howe, 1969; NIST, 2012). So the interval with limits $\bar{X} \pm kS$ is a 95%-content tolerance interval for a future measurement result with level of confidence approximately 99%.

The measurements are made and the realized values of $\bar{X}$ and S^2 are $\bar{x}$ and s^2. Thus, the interval $[\bar{x} - ks, \bar{x} + ks]$ is the realization of a 95% tolerance interval for the next measurement result with level of confidence approximately 99%. In the absence of external information, we can be approximately 99% sure that 95% of potential measurement results will lie in this interval.

In many applied contexts the title 'tolerance interval' will be unhelpful because the term 'tolerance' will have a different meaning. An alternative title could be 'finite-sample probability interval', which seems suitable because the interval acts like a probability interval but is generated from a finite set of data.

If measurement uncertainty is seen as prediction uncertainty then the concepts of a probability interval and a tolerance interval may be applicable. However, this seems unlikely to be the case. So we shall now abandon this focus on the *prediction* of values and instead consider the idea that measurement involves the *attribution* of one or more values to the measurand.

A distribution of reasonableness?

As noted, measurement is a process of obtaining one or more values that 'can reasonably be attributed to' a quantity (VIM, 2012, clause 2.1). If the implied act of reasonable attribution is not to be understood as an act of estimation or prediction then determining the uncertainty of measurement must involve quantifying how 'reasonable' it is for different values to be given to the measurand. Let us see where this idea takes us and whether we are led to a consideration of probability.

Suppose that it would only be reasonable to attribute the measurand a value in some known interval $[a, b]$, and that it would be more reasonable to attribute it some values in this interval than others. Suppose also that the 'degree of reasonableness' can be quantified in relative terms, so that it is, say, three times as reasonable to give the measurand the value x_2 as the value x_1. Further, suppose that this idea can be extended to small intervals, so that it is three times as reasonable to give the measurand a value in the small interval $[x_2, x_2 + dx]$ as a value in the interval $[x_1, x_1 + dx]$. Apart from a scale factor, this permits the definition of a function of reasonableness $r(z)$ for all values in $[a, b]$, where $r(z) = 0$ for any value z that would be completely unreasonable.

Next, suppose that it is completely unreasonable to avoid giving the measurand any value. Then we might argue that the integral over the range $\int r(z)\, dz$ must take a fixed value, which we can set to 1. The function $r(z)$ becomes a positive function integrating to 1, and so it has the appearance of a probability density function, as in Figure 12.4. However, this does not mean that it is a probability density function. For it to be a probability density function, there must be the concepts of a variate and a hypothesis whose truth is unknown, and without a target value the identity of this hypothesis is unclear. Nor does it mean that $r(z)$ can necessarily be treated like a probability density function. For example suppose we are considering attributing

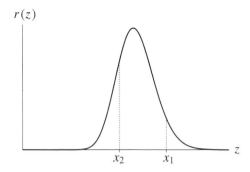

Figure 12.4 A distribution showing the relative reasonableness of attributing the value z to a measurand.

our measurement result x to the measurand. In effect, it has just been concluded that the degree of reasonableness of attributing a value between $x - \delta$ and $x + \delta$ to the measurand is $\int_{x-\delta}^{x+\delta} r(z)\,dz$, which is infinitesimal when δ is infinitesimal. So, by this logic, we are considering doing something that is completely unreasonable!

The point being made is that it would be presumptuous to regard a 'reasonableness distribution' as a probability distribution. If 'reasonableness' is the underlying concept in the matter of attributing values to quantities in measurement then what is required in complex measurement situations is a calculus of reasonableness not a calculus of probability! It is not evident that the full logic of probability would be applicable with these reasonableness distributions.

Comments

In Section 1.2, it was argued that if there is no target value then measurement is not an act of estimation but an act of declaration. The value of the measurand then refers to something that is known and is declared or created – not unknown, unknowable and estimated. Also, there is no obvious event or hypothesis about which we can make a statement of probability, in which case familiar methods for the evaluation of measurement uncertainty must be reexamined.

In this section, we have considered other implications of rejecting the idea of a target value and have sought to construct a corresponding concept of 'measurement uncertainty'. In the absence of a target value, that term seems to be a misnomer. On one hand, if measurement is the reasonable attribution of a single value to the measurand then perhaps our uncertainty or doubt is about what value to attribute. However, the value attributed will be x, the identity of which we are in no doubt! On the other hand, if measurement is the reasonable attribution of an interval of values to the measurand then this does not seem to be an interval of doubt but an interval of flexibility!

Such comments might be criticized for being overly concerned with the meanings of words in the English language. However, I believe that many of the problems of communication and standardization that we face are linked to different understandings of common words like 'uncertainty', 'probability' and 'value'. When we write of the measurand possessing a value, do we mean a known value attributed and, in effect, created in a measurement, or do we mean some unknown value to be estimated in the measurement?

13

An assessment of objective Bayesian statistics

Scientists who are dissatisfied with frequentist statistical methods seem to be turning to objective Bayesian statistics in matters of data analysis. In measurement science, objective Bayesian methodology has been proposed for uncertainty evaluation (e.g. Elster (2000), Lira (2002), JCGM (2008b, 2011)) and for the analysis of inter-laboratory comparisons (e.g. Elster and Toman (2010), Shirono *et al.* (2010)). It seems appropriate that such methodology be subject to scrutiny, so this chapter presents a criticism of objective Bayesian methodology that is much more extensive than the brief comments given in Section 12.3. The interested reader will find additional ideas discussed by O'Hagan (1994, Sections 5.32–5.49).

We will refer many times to Definition 3.1, which is repeated here for convenience.

> **Definition 3.1 (repeated)** The truth or falsehood of a hypothesis H will be determined. The probability that a consistent gambler attributes to H is the maximum amount he or she would pay in order to receive 1 unit if H is found to be true. This probability is the degree of belief of the gambler in the truth of H.

In objective Bayesian statistics the idea of personal belief is discarded in favour of an idea of rational belief based on notions of information. This is seen in the following introductory statement of Sivia (1996, pp. 9,10):

The popular argument goes that if a probability represents a degree-of-belief then it must be subjective because my belief could be different from yours. The Bayesian view is that a probability does indeed represent how much we believe that something is true, but that this belief should be based on all the relevant information available. ... As Jaynes has pointed out, objectivity demands only that two people having the same information should assign the same probability; this principle has played a key role in the modern development of the (objective) Bayesian approach.

We see that, for Sivia and others, probability is strength of belief and that belief is to follow from information. The quantification of information therefore leads to the assessment of probability.

On inspection, this quotation raises a number of issues.

1. The quotation seems to suggest that a subjective probability assignment is *not* based on all the relevant information available. Can we not imagine the subjective Bayesian statistician responding 'But my belief *is* based on all the information that I have now and have had in the past. This information has created my beliefs'?
2. The phrase 'two people having the same information should assign the same probability' only has import if it is realistic for two people to actually have the same information. This perhaps places limits on the nature of 'information'. We must ask what is meant by 'information', and what kinds of information permit the formation of a unique probability distribution.
3. The claim that 'objectivity demands ... that two people having the same information should assign the same probability' only holds true if probability has an objective meaning. Some agreed and unambiguous definition of probability is required. If Definition 3.1 (which describes personal belief) is not adequate then an alternative definition must be given, and the two definitions must not be contradictory where Definition 3.1 is applicable.
4. Also, the idea that two people with the same information should assign the same probability only follows if the act of assigning a probability to the hypothesis or event is legitimate in the first place, and this should not be presumed.

These points are concerned with ideas of 'information' and 'probability'. They relate to the various strands of criticism of objective Bayesian statistics now given. We begin by focusing on the idea of using a probability distribution to represent the state of ignorance and on the use of 'improper' prior distributions, which are often employed for this purpose. Analysis shows that using such distributions admits the possibility of drawing illogical conclusions. Then it is argued that the concept of information on which many choices of prior distribution are based does not apply with continuous random variables, such as would be relevant in a typical measurement problem. With regard to such variables, 'information' is not defined in a way that properly represents what people typically mean by that word. Following this, we give our most serious criticism of objective Bayesian statistics, which is that various probabilities involved do not clearly satisfy Definition 3.1 or any other *practical* definition of probability. Finally, it is suggested that the current popularity of objective Bayesian statistics might be based on an unjustified presupposition that it is always meaningful to make a probability statement.

13.1 Ignorance and coherence

The logical integrity and 'coherence' of the subjective Bayesian approach are often emphasized by supporters of Bayesian methods. By default, this integrity might also be used to support an objective Bayesian view. However, objective Bayesian statisticians are forced to confront the problem of representing prior ignorance about a quantity using a prior probability distribution, and various methods that have been proposed for this exhibit internal inconsistencies.

A lack of invariance to reparametrization

Let f_1 be a simple function. Suppose we wish to estimate the value of $f_1(\theta)$ when we know nothing about θ except that it lies between non-negative limits $\theta_{\min}$ and $\theta_{\max}$. A Bayesian analysis requires that θ be given a prior probability distribution. Many would regard the best choice of prior distribution to then be the continuous uniform distribution $U(\theta_{\min}, \theta_{\max})$. However, the same measurement problem can be formulated as the estimation of $f_2(\lambda)$ with $f_2(z) = f_1(\exp z)$ and $\lambda = \log \theta$. All we know about λ is that $\lambda_{\min} \leq \lambda \leq \lambda_{\max}$ with $\lambda_{\min} = \log \theta_{\min}$ and $\lambda_{\max} = \log \theta_{\max}$. So it must also be reasonable to take the prior probability distribution of λ to be uniform between these transformed limits. However, we cannot attribute uniform distributions to both θ and λ because these parameters are related in a known non-linear way. This 'lack of invariance under reparametrization' arises with all monotonic non-linear transformations. It proves that ignorance has no unique representation. Our choice of the logarithmic transformation is relevant because many quantities are measured on logarithmic scales.

The use of improper prior distributions

The requirement in objective Bayesian statistics that the state of ignorance be described quantitatively is associated with the frequent use of *improper* prior distributions, which are distributions where the density function is not integrable over the domain. As we shall see, the use of such prior distributions introduces the possibility of illogical or incoherent results. (The use of an improper distribution is admitted in objective Bayesian statistics because, in general, inference occurs only *after* observation of the data, and a proper posterior distribution can sometimes be obtained even if the prior distribution is improper.)

Because improper density functions do not integrate to unity, they do not directly provide statements of actual probability. For example consider an objective Bayesian analysis involving the unknown mean and standard deviation of the normal distribution from which we will draw a sample. To express prior ignorance

13.1 Ignorance and coherence

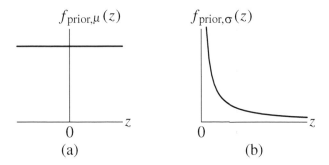

Figure 13.1 Two prior distributions often used in objective Bayesian statistics: (a) the constant prior density function $f_{\text{prior},\mu}(z) = \text{constant}$ of (13.1); (b) the hyperbolic prior density function $f_{\text{prior},\sigma}(z) \propto 1/z$ of (13.2)

about the values of these parameters, the mean μ, being a parameter that only affects location, will typically be given the uniform prior density function

$$f_{\text{prior},\mu}(z) = \text{constant}, \qquad -\infty < z < \infty, \qquad (13.1)$$

while the standard deviation σ, being a parameter that only affects scale, will typically be given the hyperbolic prior density function[1]

$$f_{\text{prior},\sigma}(z) \propto 1/z, \qquad 0 \leq z < \infty. \qquad (13.2)$$

These density functions are depicted in Figure 13.1. The prior probability attributed to the hypothesis '$\mu < 200$', say, which would be

$$\frac{\int_{-\infty}^{200} \text{constant}\, dz}{\int_{-\infty}^{\infty} \text{constant}\, dz},$$

cannot be stated. Similarly, the prior probability attributed to the hypothesis '$\sigma < 1$', say, which would be

$$\frac{\int_0^1 1/z\, dz}{\int_0^\infty 1/z\, dz},$$

cannot be stated.

Because improper distributions do not admit statements of actual probability, they cannot properly represent strength of belief as defined in Definition 3.1. Therefore calculus with these distributions does not necessarily exhibit the coherence associated with the beliefs of the 'rational agent' or 'consistent gambler'. Several examples of the logical frailty of an analysis with improper prior distributions

---

[1] Using (13.2) is equivalent to giving σ^2 the prior density $f_{\text{prior},\sigma^2}(z) \propto 1/z$ for $z \geq 0$ and $\log \sigma$ the prior density $f_{\text{prior},\log \sigma}(z) = \text{constant}$ for $-\infty < z < \infty$.

follow. The first involves a model and dataset that are realistic in a measurement context.

Example 13.1 Let the value of the measurand θ be directly estimated in n independent measurements. Suppose that each measurement incurs random error from a normal distribution with mean 0 and some unknown variance σ^2, and suppose that the raw response is then displayed on a digital meter in which it has been rounded down to the nearest unit. Therefore, for the ith measurement, the model is

$$e_{\text{raw},i} \leftarrow \text{N}(0, \sigma^2),$$
$$x_i = \lfloor \theta + e_{\text{raw},i} \rfloor$$

where $\lfloor \cdot \rfloor$ means rounding down to the nearest unit.

The estimation of θ in a Bayesian analysis would usually involve the prior density functions (13.1) and (13.2) held independently. By a limiting argument involving a sequence of proper posterior distributions, Elster (2000) derives the corresponding posterior distribution of θ when $x_1 = \ldots = x_n = x$ to be the uniform distribution with lower limit x and range one unit. This result tells us that when all the n results are equal to x, say, we become *certain* that θ lies in the interval $[x, x+1]$; the posterior probability that θ lies outside this interval is zero. However, there will obviously be situations where all the data happen to fall in a single interval different from the interval containing θ, especially when n is small (Willink, 2007b). Prior ignorance when expressed in the prior density functions (13.1) and (13.2) has become unjustified certainty!

The second example relates to the measurement of a non-negative target value in the presence of random error. It involves the idea of a *predictive distribution*, which is a probability distribution calculated for a future datum.

Example 13.2 Consider measuring a target value θ known to be non-negative. Suppose we know that each time a measurement is made the error is independently drawn from the standard normal distribution $\text{N}(0, 1)$. Let X_1 and X_2 be the variates for two measurement results to be obtained, so $X_1 \sim \text{N}(\theta, 1)$ and $X_2 \sim \text{N}(\theta, 1)$. These are independent and identically distributed variates on a continuous scale, so $\Pr(X_2 > X_1) = 0.5$.

Many who favour an objective Bayesian approach would then see the density function

$$f_{\text{prior},\theta}(z) = \text{constant}, \qquad z \geq 0, \qquad (13.3)$$

as the appropriate substitute for (13.1). Suppose this is chosen as the prior density function of θ. The first measurement is made and the result is x_1. The posterior distribution of θ is a normal distribution with mean x_1 truncated below at 0 and renormalized, as in Figure 13.2(a) for $x_1 > 0$ and Figure 13.2(b) for $x_1 < 0$.

13.1 Ignorance and coherence

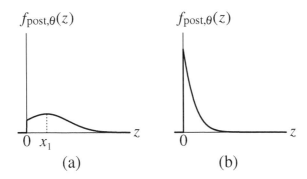

Figure 13.2 Posterior distributions of θ obtained after observing the value x_1 taken by X_1: (a) $x_1 > 0$, (b) $x_1 < 0$.

The probability distribution subsequently derived for X_2, which is the *posterior predictive distribution* of the next measurement result, is the distribution of the sum of independent variates with this truncated distribution and the standard normal distribution. Using the sense of asymmetry about x_1 that is seen in both parts of Figure 13.2, we conclude that most of the probability content of this predictive distribution lies above the value x_1. Thus, the posterior predictive probability that X_2 takes a value greater than x_1 is $\text{Pr}_{\text{post}}(X_2 > x_1) > 0.5$. (In fact, it can be shown that $\text{Pr}_{\text{post}}(X_2 > x_1) = 1 - \Phi(x_1)/2$.) So, whatever the value taken by X_1, the objective Bayesian statistician then considers X_2 to have probability greater than one-half of taking a larger value. This is inconsistent with the obvious truth that $\text{Pr}(X_2 > X_1) = 0.5$ at the outset.

The third example also considers the measurement of quantities known to be positive. It involves combining the results of measurement of two or more such quantities, such as the concentrations of different species of pollutant, in order to measure their sum.

Example 13.3 Suppose the concentration θ_1 of a pollutant is measured using a process with normal error having known standard deviation σ. In an objective Bayesian analysis using (13.3) as the prior distribution for θ_1, the posterior distribution will be a truncated normal distribution, as seen in Figure 13.2. Measurement of the concentration θ_2 of a second pollutant using a process with the same accuracy will give another truncated normal distribution. The distribution calculated for the total level of pollution $\theta = \theta_1 + \theta_2$ will be obtained by convolving the two individual distributions.

Suppose, for example, that the distributions obtained for θ_1 and θ_2 are those depicted in Figures 13.2(a) and 13.2(b) respectively. The horizontal scale is the same in both. Figure 13.3(a) shows the corresponding distribution of θ, with the vertical scale being expanded by a factor of 3 for clarity. The probability density function

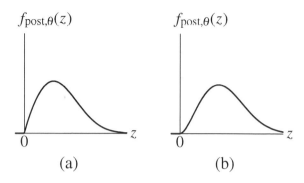

Figure 13.3 Distributions calculated for the sums of quantities attributed the distributions shown in Figure 13.2: (a) $\theta = \theta_1 + \theta_2$ where θ_1 and θ_2 have the distributions shown in Figure 13.2(a) and Figure 13.2(b) respectively; (b) $\theta = \theta_1 + \theta_2 + \theta_3$ where θ_3 also has the distribution shown in Figure 13.2(b).

$f_{\text{post},\theta}(z)$ is continuous and is zero at the origin, so the posterior probability being given to the idea that $\theta \approx 0$ is negligible, and the shortest 95% credible interval for θ will not contain the origin. This would be the case whatever the values of x_1 and x_2. So if θ_1 and θ_2 are negligible, which would be hoped for with concentrations of pollutants, the 95% uncertainty interval calculated for θ will always lie fully above θ. The Bayesian methodology will then have done a poor job of estimating θ.

It might be argued that, because there is no 'tail' at the lower end of the posterior distribution in Figure 13.3(a), the credible interval containing the lowest 95% of the distribution should be used instead. However, if the measurand were the sum of three or more components then $f'_{\text{post},\theta}(0) = 0$ as well as $f_{\text{post},\theta}(0) = 0$, and this provides the posterior density function with the makings of a tail. (For a proof of this result, see Willink (2010b).) For example Figure 13.3(b) shows the distribution obtained for $\theta = \theta_1 + \theta_2 + \theta_3$ when the posterior distribution of θ_3 is also the distribution seen in Figure 13.2(b).

So the results of measuring the sum of quantities given independent flat prior distributions on the positive real axis can be highly misleading.

There is also another type of inconsistency existing with the typical objective Bayesian analysis in this example situation. Imagine that a different experimenter has a one-step technique of measuring the total concentration $\theta = \theta_1 + \theta_2$ with double the variance. So the experimental performance of this method and the two-step technique of the first scientist are identical. The new experimenter would attribute θ a flat prior distribution and would obtain a different posterior distribution for θ (this distribution being a single truncated normal distribution). So the results of the objective Bayesian analysis depend entirely on whether θ is approached as the sum of two separate components or as a single overall concentration. This logical inconsistency seems unacceptable.

13.1 Ignorance and coherence

These examples have shown that the use of improper prior distributions admits incoherence. Thus, the traditional objective Bayesian view lacks the theoretical justification that is enjoyed in subjective Bayesian statistics and that is sometimes, by implication, claimed for all of Bayesian statistics.

The representation of ignorance

The following examples give evidence of other kinds of logical inconsistency admitted when using improper prior distributions. A common thread between these three examples is the idea that these distributions are used to represent prior ignorance. The paradoxical behaviour observed is related to the idea of expressing ignorance rather than to the improper nature of the distributions.

Like Example 13.2, the first example relates to the measurement of a single non-negative quantity in the presence of random error.

Example 13.4 Consider measuring a quantity θ known to be non-negative. Suppose that each time we use the measurement technique the error is independently drawn from the normal distribution $N(0, \sigma^2)$ with σ unknown. A number n of measurements are made and the results are averaged to form the overall measurement result x. The corresponding sample standard deviation is s.

The objective Bayesian statistician is likely to advocate the use of the prior density functions (13.2) and (13.3) to describe prior ignorance about σ and θ. The posterior distribution calculated for θ will be the distribution of $x + s/\sqrt{n}T_{n-1}$, truncated below at zero and renormalized. This distribution is seen in Figure 13.4 for $n = 2$, for a certain value of $s/\sqrt{n}$, and for different values of x.

When $x = s/\sqrt{n}$ the 95th percentile of the posterior distribution of θ is $9.45s/\sqrt{n}$, but when $x = 0$ the 95th percentile of the posterior distribution of θ is $12.71s/\sqrt{n}$, which is larger! This result is enlightening: the objective Bayesian statistician is saying, 'I know nothing about θ except that it is positive. Suppose I make $n = 2$ measurements and the sample standard deviation is the non-zero figure s. If the sample mean is $x = s/\sqrt{n}$ then I will be 95% sure that $\theta < 9.45s/\sqrt{n}$ but if the sample mean is zero then I will be 95% sure that $\theta < 12.71s/\sqrt{n}$.'

Several other points can be made relating to this measurement situation.

1. The sample mean x can conceivably be negative when θ is small compared to σ. As x decreases from $+1.455s/\sqrt{n}$ and becomes increasingly negative the posterior 95th percentile increases from $9.29s/\sqrt{n}$ without bound!
2. The basic effect that a posterior quantile can increase as x decreases occurs with all finite values of $n \geq 2$. This is due to the fact that the probability density function of T_ν is log-convex in the upper tail beyond the point $\sqrt{\nu}$.
3. This basic effect exists even when (13.3) is replaced by a proper density function!

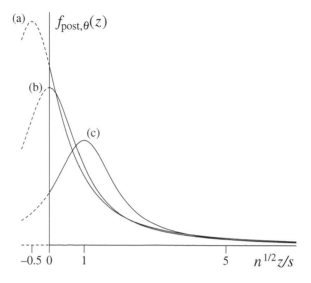

Figure 13.4 Posterior density functions of θ obtained using a sample of size $n = 2$ with sample mean x and a certain sample standard deviation s: (a) $x = -0.8s/\sqrt{n}$; (b) $x = 0$; (c) $x = s/\sqrt{n}$.

4. Suppose we consider the *subjective* Bayesian problem of estimating θ from a single observation x with error $x - \theta$ drawn from a known scaled t distribution. Here the only unknown, θ, will be given a proper probability distribution. The effect is still observed, but the claim of prior ignorance that led to the italicized statement being ridiculous is absent. The behaviour is counter-intuitive rather than unacceptable.

5. Now consider the objective Bayesian problem of estimating θ when the error is drawn from a known scaled t distribution. The effect is observed whatever prior distribution is chosen for θ (provided that the posterior distribution is proper). Is this not another proof that ignorance admits no consistent representation?

In the second example, the inconsistency is related to the idea that if prior ignorance about a quantity admits a unique mathematical representation then this representation cannot depend on the way in which we eventually choose to study the quantity.

Example 13.5 An experimenter envisages the measurement of a quantity θ. Two methods are available. The first method is known to incur additive error having a normal distribution with mean zero and some fixed variance. The corresponding model is

$$X_1 = \theta + E_1, \qquad E_1 \sim N(0, \sigma_1^2),$$

13.1 Ignorance and coherence

with σ_1^2 being either known or unknown. Equivalently, $X_1 \sim N(\theta, \sigma_1^2)$. An objective Bayesian statistician would identify θ as behaving like a location parameter in the distribution of responses and so would advocate the use of a uniform prior density function for θ. The second method is known to incur multiplicative error having a normal distribution with mean zero and some fixed variance σ_2^2. The corresponding model is

$$X_2 = \theta(1 + E_2), \qquad E_2 \sim N(0, \sigma_2^2).$$

The objective Bayesian statistician would then see θ behaving as a scale parameter and so would choose the hyperbolic density $f(z) \propto 1/z$ for θ.

So the prior distributions advocated in objective Bayesian statistics differ in the two cases. But how can belief about θ prior to the experiment be influenced by the method that will subsequently be used to approximate θ? The foundational Bayesian idea that the prior distribution describes prior belief is not being honoured.

In the last of our examples, the inconsistency relates to the idea that if there is a unique mathematical representation of ignorance about a quantity then this representation cannot depend on whether we are interested in studying the quantity by itself or simultaneously with other quantities.

Example 13.6 Consider the measurement of a p-dimensional quantity expressed as a column vector $\boldsymbol{\theta} = (\theta_1, \ldots, \theta_p)'$. Suppose that the result of any single measurement can be assumed to be drawn from the multivariate normal distribution with mean $\boldsymbol{\theta}$ and unknown $p \times p$ covariance matrix $\boldsymbol{\Sigma}$. Geisser and Cornfield (1963) discuss Bayesian estimation of $\boldsymbol{\theta}$ using the prior density function

$$f_{\boldsymbol{\theta}, \boldsymbol{\Sigma}^{-1}}(\boldsymbol{\Lambda}) \propto |\boldsymbol{\Lambda}|^{k/2}, \tag{13.4}$$

which they indicate is a form of prior density function that can be justified using an objective Bayesian criterion. Suppose that there are n independent measurements of $\boldsymbol{\theta}$, that $\bar{\mathbf{x}}$ is the observed sample mean vector and that $\mathbf{s}$ is the usual sample covariance matrix (i.e. the matrix obtained using the denominator $n - 1$). Geisser and Cornfield show that the corresponding posterior probability density function of $\boldsymbol{\theta}$ is

$$f_{\boldsymbol{\theta}}(\boldsymbol{\xi}) \propto \left\{ 1 + \frac{1}{n-1}(\boldsymbol{\xi} - \bar{\mathbf{x}})' \left(\frac{\mathbf{s}}{n}\right)^{-1} (\boldsymbol{\xi} - \bar{\mathbf{x}}) \right\}^{-(n+p+1-k)/2}, \tag{13.5}$$

with $\boldsymbol{\xi}$ being a dummy variable.

As noted by Geisser and Cornfield, when $k = 1 + p$ (and $n > p$), the posterior distribution of the quantity $(\boldsymbol{\theta} - \bar{\mathbf{x}})'(\mathbf{s}/n)^{-1}(\boldsymbol{\theta} - \bar{\mathbf{x}})$ is the distribution of Hotelling's T^2 with parameters p and $n - 1$. This relationship means that when $k = 1 + p$ the 95% hyperellipsoidal credible region centred on $\bar{\mathbf{x}}$ is the same as the realized 95% confidence region calculated in Section 10.1. Thus the prior distribution with $k = 1 + p$ in (13.4) is known as the 'matching prior' when estimating a p-dimensional multivariate-normal mean vector such as $\boldsymbol{\theta}$. The idea of choosing a prior distribution with this property of matching the numerical frequentist result might be appealing, and so many objective Bayesian statisticians would use this prior distribution.

Geisser and Cornfield then show that, for every non-zero vector $\boldsymbol{a}$ of length p, the scalar quantity $\boldsymbol{a}'\boldsymbol{\theta}$ has the distribution of $\boldsymbol{a}'\bar{\mathbf{x}} + (n-1)/(n-p)\sqrt{\boldsymbol{a}'\mathbf{s}\boldsymbol{a}/n}\, T_{n-p}$ where T_ν is a random variable with Student's t distribution with ν degrees of freedom. So the marginal distribution of θ_j, which is obtained when $\boldsymbol{a}$ has jth element equal to 1 and all other elements zero, is the distribution of a variable $\bar{x}_j + (n-1)/(n-p) \times s_j/\sqrt{n}\, T_{n-p}$, where $\bar{x}_j$ and s_j^2 are the observed sample mean and observed sample variance for the n results of measuring θ_j. But, when $p > 1$, this distribution differs from the distribution that would be the output of an objective Bayesian analysis if θ_j had been the quantity of interest, namely the distribution of $\bar{x}_j + s_j/\sqrt{n}\, T_{n-1}$. As we have seen, that is the posterior distribution we would have obtained using the prior distribution (13.1)–(13.2), which is the matching prior in the estimation of a univariate normal mean such as θ_j. (This difference can be observed in distributions recommended for use in the first and second supplements to the *Guide* (JCGM, 2008b, clause 6.4.9; 2011, clause 5.3.2), which address univariate and multivariate problems respectively.)

How are we to understand this difference? Should the distribution that describes our resulting belief about θ_j depend on whether $\boldsymbol{\theta}$ or θ_j was the primary quantity of interest? It might be argued that the data obtained when measuring the other elements of $\boldsymbol{\theta}$ could affect our belief about θ_j, but those data do not appear at all in the expression $\bar{x}_j + s_j/\sqrt{n}\, T_{n-1}$. Anyway, if the existence of the other data is at all relevant to the estimation of θ_j then it would have been expected to improve our knowledge and so lead to a posterior distribution of θ_j with a reduced variance. But through the involvement of the additional scale factor and a smaller number of degrees of freedom, the variance is actually greater. If this difference is not to be regarded as an inconsistency then we must conclude that the construction of a marginal distribution is a meaningless operation in objective Bayesian statistics!

13.2 Information and entropy

Often the claim is made that the Bayesian paradigm can incorporate information not utilized in frequentist statistics. But what kind of information is being referred to, and is this type of information sufficient for an acceptable prior probability distribution to be formed? One situation frequently cited is where the parameter to be estimated, θ, is known from physical considerations to have a certain bound. Suppose we know that $\theta \geq 0$. This knowledge enables us to attribute zero prior probability density to θ at values below zero, but it can hardly be said to constitute information from which a prior probability density function can be constructed at positive values. Although there is certainly some prior information about θ in this situation, it is one discrete piece of information, not the continuum of information

13.2 Information and entropy

that would be required to attribute θ a (proper) probability density at each point in the continuous domain $[0, \infty)$. Thus, information is not always of a type that is adequate for a Bayesian analysis with proper prior distributions – while it is only an *absence* of information that would seem to encourage the use of an improper prior distribution. (Incidentally, a thoughtful frequentist analysis would make use of the information that $\theta \geq 0$. This information would be used to raise any limit to zero. This step would be a natural part of the frequentist procedure because it cannot reduce the success rate of the procedure below the advertised figure of 95%. The frequentist statistician is able to make use of this information without having to make any other assumptions.)

Sometimes the choice of prior distribution is made by appealing to a definition of 'information' associated with work of Shannon (1948) in communication theory. Let S be a categorical variate with n categories or 'symbols', $s_1, \ldots, s_n$, and with $\Pr(S = s_i) = p_i$. Consider assigning codewords such as 0, 10, 110, 1110 to these symbols and to groups of these symbols. Shannon (1948, Theorem 9) showed that (i) no assignment of codewords can allow an infinitely long series of independent realizations of S to be transmitted in fewer than $-\sum_{i=1}^{n} p_i \log p_i$ bits per symbol and that (ii) this bound can be attained in theory. Therefore, as k tends to infinity, the amount of information contained per realization in k realizations of S becomes the same as the amount of information contained per digit in a certain optimal binary sequence of length $k \sum_{i=1}^{n} p_i \log p_i$. This *source-coding theorem* justifies equating the quantity

$$H(S) = -\sum_{i=1}^{n} p_i \log p_i \qquad (13.6)$$

with a rate of gain of information from repeated observation of S. This quantity is called the '(mathematical) entropy' of S.

From this idea that entropy is inversely related to information rate comes the idea that the probability distribution maximizing the entropy of a variate subject to any known constraints is the distribution that is least informative about the value taken by that variate. This 'principle of maximum entropy' might then be invoked when a prior probability distribution has to be attributed to an unknown quantity in a Bayesian analysis and when this choice of distribution is intended to have minimal effect on the results obtained. In so doing, the concept of *information rate* is subtly being rendered as a concept of *information* itself.

The original context of Shannon was that of repeated observation of a categorical variable. He was able to obtain a similar result for a continuous numerical variable but only by assuming a certain loss function to describe information fidelity (Shannon, 1948, Theorem 22). However, Jaynes (1982) extrapolates to state

The general Principle of Maximum Entropy is applicable to any problem of inference with a well-defined hypothesis space and noiseless but incomplete data, whether or not it involves a repetitive situation ....

I have found no clear support for this extrapolation in that paper of Jaynes or in earlier relevant papers (Jaynes, 1957, 1968). In fact, as now argued, the idea that entropy relates to 'information' does not seem to hold with single realizations of numerical variables.

The original definition of entropy in this context, which is (13.6), describes a property of a categorical variate. Similarly, the entropy of a discrete numerical variate X with probability mass function $\Pr(X = i) = p_i$ is defined as

$$H(X) = - \sum_{i=-\infty}^{\infty} p_i \log p_i, \qquad (13.7)$$

while the (differential) entropy of a continuous variate X with probability density function $f_X(z)$ is defined as

$$H(X) = - \int_{-\infty}^{\infty} f_X(z) \log f_X(z)\, dz. \qquad (13.8)$$

Although we might generalize the definition of entropy beyond categorical variates in this way, it does not follow that the quantities in (13.7) and (13.8) can automatically be identified with information or information rate. The relevant concepts of information with numerical variates will be seen to differ from those with categorical variables.

With categorical variables information is found in the ideas of equality and inequality. For example the statements '$s_1 = s_2$' and '$s_3 \neq s_4$' are meaningful. But with discrete numerical variables, information is also found in the ideas of order and distance. For example if n is the number of decay events observed in some experiment then the statements '$n > 10$' and '$n \approx 400$' are informative statements of order and closeness. Moreover, with continuous numerical variables, information *only* relates to order and distance because it is impossible for two non-identical variables on a continuous scale to take exactly the same value. Thus, order and distance are important sources of information with numerical variables and are the only sources of information with continuous variables. However, the concepts of order and distance are absent when discussing categorical variables. For example the statements 'red $\approx$ blue' and 'Portugal > New Zealand' have no meaning. In fact, reordering and rearranging the domain in (13.7) and (13.8) – which grossly affects both order and distance between potential values of the numerical variates – makes no change to the entropies; see Figure 13.5.

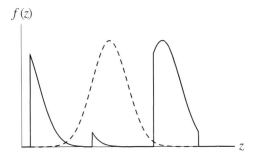

Figure 13.5 Two continuous distributions with equal entropy as defined in (13.8). Dashed curve: a normal probability density function; solid curve: a density function formed by reordering and rearranging sections of the solid curve.

So the ideas of information that apply with numerical variates, and especially with continuous numerical variates, are quite different from the kind of information that can be quantified in terms of entropy. Moreover, any idea of information in a single realization of such a variate does not involve the concept of repetition or information *rate*, which was essential in Shannon's analysis. Therefore, although an objective Bayesian statistician might invoke the 'principle of maximum entropy' when attributing a prior distribution to an unknown fixed quantity, there appears to be little justification for doing so.[2]

Reference analysis

The field of so-called 'information theory' grew out of Shannon's results about entropy. Concepts from this field have been used in the development of *reference analysis*, which might be seen as an approach within the field of objective Bayesian statistics. This involves the use of prior distributions that are intended not to describe personal belief (as in Definition 3.1) but to act as 'formal *consensus* prior functions to be used as standards for scientific communication' (Bernardo (2005b) with his italics). From within an acceptable class of distributions, the prior distribution for an unknown parameter θ is chosen so that (assuming the sampling model is correct) the 'information' gained about θ by subsequent observation of the data is maximized. Once again, there is a dependence on a mathematical concept of information with dubious relevance to continuous variables. Given that the interpretation of the prior distribution is not identified with a concept of belief, there is also doubt about the meaning of the results.

---

[2] For other criticism of the concept of identifying information with entropy, see Kåhre (2002).

13.3 Meaning and communication

This book has sought to discuss the relevance of 'probability' to the ideas of measurement uncertainty. To do this, we have taken the reasonable steps of defining probability and examining the implications of this definition. This has resulted in Definitions 3.1 and 3.2. It has also resulted in the related claim that acceptable probability statements (about independent hypotheses) must obey a long-run interpretation, which is a claim that follows from examining the rationale of betting for positive expected gain. In the context of this chapter, this long-run interpretation is as follows: if a probability p_i is objectively attributed to the ith hypothesis in a set of independent hypotheses then, as n increases without bound, the proportion of these hypotheses that are true must be $\lim_{n \to \infty} \sum_{i=1}^{n} p_i/n$.

However, the prior probability distributions used in objective Bayesian statistics do not seem to be constructed in ways that obey Definition 3.1, and this implies that the posterior distributions obtained cannot be seen as obeying it either. Instead, prior distributions are often constructed according to rules that relate to the availability of 'information', as understood in narrow senses of the word. The concept of actual belief is at the heart of Definition 3.1, but if this concept is de-emphasized then the force and appropriateness of this definition seem diminished, and an alternative definition of probability is required. Yet it is difficult to see in objective Bayesian literature any practical alternative definition of probability. In fact, the idea that a numerical probability, say 0.95, is to have a practical meaning seems missing from almost all descriptions of objective Bayesian statistics. This omission is perhaps related to the claim that the sum of the probabilities of all outcomes is only set to unity by convention (see Sivia (1996), p. 3).

Presumably most objective Bayesian statisticians would accept Definition 3.1 and many would accept its long-run implication. Indeed, the idea that a posterior probability in a Bayesian analysis is to be consistent with a long-run success rate is upheld in the quotation from Bayarri and Berger (2004) given in Section 12.3. A similar idea can be seen in work of Bernardo (2005a), who has suggested that objective Bayesian methods (in the context of teaching statistics, at least) could be evaluated according to frequentist criteria.[3] However, often the idea that probabilities relate to 'success rates' or 'truth rates' seems absent when objective Bayesian rules are applied. And we must ask what evidence exists to show that the statistician would behave in the manner implied in Definition 3.1 when the probability in question is constructed using such rules. Our claim, therefore, is that the outputs of many objective Bayesian analyses lack practical meaning.

---

[3] Interestingly, the philosophical distance between this suggestion of Bernardo and the frequentist understanding is much less than the distance between this suggestion and the subjective Bayesian view.

13.4 A presupposition?

This chapter has contained severe criticism of some of the ideas of objective Bayesian statistics. Yet many people seem to be choosing this approach to statistical inference. We should ask ourselves why this is the case. One possibility is that the Bayesian paradigm claims that direct probability statements can be made about unknown fixed quantities. Undoubtedly this claim is appealing, but is it scientifically acceptable? And is this claim actually a presupposition for many people? Some remarks of Fisher seem relevant here (Fisher, 1930, pp. 528, 529; Pedersen, 1978). He imagines an argument by which scientists such as Laplace and Gauss could fall into error in accepting the basic Bayesian approach, and writes:

> In fact, the argument runs something as follows: a number of useful but uncertain judgements can be expressed with exactitude in terms of probability; our judgements respecting causes or hypotheses are uncertain, therefore our rational attitude towards them is expressible in terms of probability. The assumption was almost a necessary one seeing that no other mathematical apparatus existed for dealing with uncertainties.

Fisher is commenting on an unwarranted extrapolation of the idea of probability from one set of problems to another.

In a number of simple situations, the numerical intervals obtained in an objective Bayesian analysis and a frequentist analysis are identical (or can be made identical by the choice of a matching prior). To a person focused only on outputs, but not to a philosopher, this fact might seem to support the use of objective Bayesian methods. Arguably, this person is comfortable with the numerical results obtained in the frequentist analysis but wants to think about them in a different way. He or she wants to make a direct probability statement about the measurand, not a confidence statement. The equivalence of a 95% credible interval with a realized 95% confidence interval in familiar simple cases encourages the person to believe that this is possible.

Perhaps it has been presumed that a statement of probability about an unknown fixed quantity can be scientifically meaningful. When a particular choice of prior distribution in an objective Bayesian analysis is criticized, we may hear the response 'What other should I choose? After all, I must choose *some* prior distribution!' No, that is not the case. The person is only compelled to specify a prior distribution for an unknown constant if he or she has concluded that the proper approach to data-based scientific inference is to make a direct probability statement about this constant after application of Bayes' theorem. Has that conclusion been based on a presupposition that scientific probability statements can in fact be made about constants in the first place? If such a presupposition were discovered and abandoned, the person would be freed from the obligation to construct or invent such a prior distribution.

So has it been assumed that probability statements can be made about constants and that there must be an acceptable way to do that? The author would argue that this assumption would be faulty and that the difficulty objective Bayesian statisticians have had in delivering a unified theory of inference is testimony to that. When faced with an insurmountable obstacle, we must retrace our steps until we reach a point where another path can be chosen. Arguably, those who have assumed that scientifically acceptable probability statements must exist in all experimental situations are faced with such an obstacle.

13.5 Discussion

The fundamental difference between objective and subjective Bayesian statistics relates to the choice of prior distributions. In objective Bayesian statistics, the intention seems to be to choose prior distributions that have minimal effect on the conclusions that will be drawn. As an input to Bayes' theorem, the prior distribution becomes an inconvenience rather than a meaningful ingredient of the decision-making process. Presumably, many of those who favour this approach would do away with the need for prior distributions if only they could. For example we can read of (Bernardo, 2005a):

> ... the emergence of powerful objective Bayesian methods (where the result, as in frequentist statistics, only depends on the assumed model and the observed data) ...

In this, the idea that the prior distribution plays no role as an independent entity seems regarded as a point of merit.

In summary, I concur with Feldman and Cousins (1998, p. 3874) when they suggest that scientists might take a subjective Bayesian approach to decision-making in their work but that confidence intervals are to be preferred over 'objective Bayesian' credible intervals if results are to be published in an objective way. I also regard the objective Bayesian approach as one in which scientists have sought the best of two worlds, ultimately without success. The (pre)supposition that scientific statements of probability can be made about fixed unknowns seems to be at fault.

14

Guide to the Expression of Uncertainty in Measurement

A fundamental step in the development of our subject was taken in 1980 when the Bureau International des Poids et Mesures convened a Working Group to discuss the calculation and statement of measurement uncertainty. The report from this group included 'Recommendation INC-1 (1980)', which was used as the basis for the development of the *Guide to the Expression of Uncertainty in Measurement* (the *Guide*) (BIPM *et al.*, 1995; JCGM, 2008a). The *Guide* (or the GUM as it is often known) is widely regarded as the international standard regarding the evaluation and expression of uncertainty in measurement. Consequently, it seems appropriate in this book to comment on the *Guide*.

14.1 Report and recommendation

The text of the report of the Working Group (Kaarls, 1980) makes the intention of the group clear: there is to be a unified treatment of errors associated with random observations (category A) and all other errors (category B). An error or an 'uncertainty' in either category is to be described by a standard deviation (or a variance). This is summarized in the following statements (Kaarls, 1980, p. 6, abstract, p. 7), which we have already seen in Chapter 4.

The traditional distinction between "random" and "systematic" uncertainties (or "errors", as they were often called previously) is purposely avoided here, ...

and

The new approach, which abandons the traditional distinction between "random" and "systematic" uncertainties, recommends instead the direct estimation of quantities which can be considered as valid approximations to the variances and covariances needed in the general law of "error propagation".

Also, we read:

> The essential quantities appearing in this law are the variances (and covariances) of the variables (measurements) involved, ...

which is a statement that refers to variances of measurements, not measurands.

Regrettably, Recommendation INC-1 (which is reproduced in Clause 0.7 of the *Guide*) is not as clear as the main text of the report. Central clauses in this recommendation, with their original numbering, are as follows:

> 2. The components in category A are characterized by the estimated variances, s_i^2, (or the estimated "standard deviations" s_i) and the number of degrees of freedom, v_i. Where appropriate, the estimated covariances should be given.
>
> 3. The components in category B should be characterized by quantities u_j^2, which may be considered as approximations to the corresponding variances, the existence of which is assumed. The quantities u_j^2 may be treated like variances and the quantities u_j like standard deviations. Where appropriate, the covariances should be treated in a similar way.
>
> 4. The combined uncertainty should be characterized by the numerical value obtained by applying the usual method for the combination of variances. The combined uncertainty and its components should be expressed in the form of "standard deviations".

These statements do not make clear the idea evident in the text of the report that the variances are properties of measurements, i.e. measurement results, not measurands. That is, the recommendation does not clearly indicate that the quantities given variances are error variates[1] (random variables) like $X_i - \theta_i$ not fixed quantities like θ_i. The person who reads the recommendation, but not the full report, is left in doubt (especially with regard to Type B evaluation of uncertainty) as to what entities are to be attributed standard deviations, i.e. what entities are to be considered variates.

Arguably, this omission has contributed to a problematic mixing of ideas within the *Guide*. In some parts of the *Guide* the only quantities considered to be variates are the potential results of measurements, as in frequentist statistics, but in other parts of the *Guide* it is possible to read the text in a way that allows the fixed unknown value of a quantity to be considered a variate. Thus the frequentist paradigm and the distributed-measurand concept, which are incompatible, both seem to be accepted in the same procedure.

---

[1] This is, however, implied if the correct interpretation of the figure s_i^2 in clause 2 can be assumed.

14.2 The mixing of philosophies

The *Guide*'s treatment of uncertainties in category A (i.e. Type A evaluation of uncertainty) is firmly grounded in frequentist statistics, where the only distributions involved are the distributions of potential measurement results or errors. For example Clause 4.2, Annex G, Example H.3 and Example H.5 of the *Guide* give analyses of observational data that are entirely in keeping with those in frequentist texts and make no mention of any other approach to statistics being taken. In contrast, the description of the evaluation of uncertainties in category B (i.e. Type B evaluation of uncertainty) is ambiguous. For example Clause 4.3.5 presents an entity X_i as having probability 0.5 of lying in an interval between known limits. Ambiguity arises because it is unclear if the symbol X_i is being used to represent an estimator of the quantity being measured or the quantity itself. (Clause 4.1.1 acknowledges that the same symbol is used for both.)

Indeed, some scientists seem to have concluded that Type B evaluation is based on the idea that probability distributions can be assigned to fixed unknowns – which is the distributed-measurand concept – as in the fiducial or Bayesian points of view (Kacker *et al.*, 2007; JCGM, 2008b, 2011). Yet neither the fiducial nor Bayesian approaches to inference is acknowledged in the *Guide*, and there is no mention of prior distributions, which are essential components of a Bayesian analysis.

The issue seems to be associated with the fact that Type B evaluation involves the concept of probability as degree of belief, as mentioned in Clause 3.3.5 of the *Guide*. It has been argued that Clause 4.1.6 allows the logical combination of Type A and Type B evaluations under this concept (Kacker *et al.*, 2007). But the problem remains if Type A evaluation is thought of as involving degree of belief *about a potential measurement result*, as in frequentist statistics, and Type B evaluation as involving degree of belief *about an unknown constant*, as is permitted in fiducial or Bayesian inference. Perhaps emphasis on a contrast between 'degree of belief' versus 'long-run frequency' has obscured the idea described in Section 3.1 of this book that the real difference between the competing views is about what entities are regarded as variates.

So the *Guide* might then be seen as wrongly mixing statistical paradigms. These paradigms relate to the different branches of the following question: is a quantity to be estimated interpretable as *a parameter* of a probability distribution or as *the subject* of a probability distribution? These conflicting ideas are seen side by side in Clauses 4.4.3 and 4.4.5 where we read '... is assumed to be the best estimate of the expectation μ_t of t ...' and '... the best estimate of t is its expectation $\mu_t = \ldots$'. In the first statement μ_t is being estimated, but in the second statement μ_t is known and is being used to estimate the entity t.

So the theoretical basis of the *Guide* is unclear. This issue was acknowledged by the time of publication of Supplement 1 to the *Guide* (JCGM, 2008b), which describes a numerical method of analysis based on Monte Carlo methodology. Supplement 1 advocates the Bayesian point of view and supports the use of maximum-entropy distributions for fixed quantities, but in doing so it moves further from the intentions of the Working Group of 1980. It is clear that the Working Group intended the interpretation of uncertainties in category B to be brought into line with the classical interpretation of uncertainties in category A. However, the step taken in Supplement 1 is to make the treatment of category A uncertainties theoretically consistent with the particular treatment of uncertainty that the authors of Supplement 1 advocate, one that is said to be 'generally consistent with the concepts underlying the GUM' (JCGM, 2008b, Introduction).

The problem is exacerbated by the fact that neither the *Guide* itself nor Supplement 1 specify any practical meaning for the uncertainty intervals that they describe. The *Guide* implies in various clauses relating to 'expanded uncertainty' that a 95% interval of measurement uncertainty calculated using its procedure is *an interval about the result of a measurement that may be expected to encompass 95% of the distribution of values that could reasonably be attributed to the measurand* (e.g. Clause 2.3.5), while Supplement 1 advocates the calculation of a coverage interval for the value of the measurand, with a 95% coverage interval being implied in its Clause 3.1.2 to be *an interval containing the value of the measurand with 95% probability, based on the information available*. Neither document acknowledges success rate as a relevant concept, and neither offers any practical performance requirement – measurable or unmeasurable – for an uncertainty evaluation procedure.

So the *Guide* appears to adopt two different and incompatible approaches to probability, these being the frequentist approach in Type A evaluation of uncertainty and a distributed-measurand approach at some places in its description of Type B evaluation. There seems to be no unique correct interpretation of the *Guide*, and readers should be wary of statements to the contrary made many years after its initial publication. One beneficial step would be to identify factors that have contributed to this confusion with the view to eliminating these factors from future versions of the *Guide*.

Contributing factors

Undoubtedly some of the confusion has been caused by a lack of clarity in notation and language within the *Guide*. With regard to notation, the *Guide* states in Clause 4.1.1 that it uses the same symbol for the physical quantity (the measurand) and for the random variable that represents the possible outcome of an observation of that quantity. The entity that is estimated is not distinguished from the entity that does the estimation![2] It has been stated that revision of the *Guide* will address this (Bich, 2008).

With regard to language, one difficulty can be observed with the term 'quantity'. Clauses 4.1.1 and 4.1.2 indicate that the measurand Y is definable by an equation $Y = f(X_1, \ldots, X_n)$, where $X_1, \ldots, X_n$ are input quantities that themselves may have been measurands in earlier measurements. Are these input quantities 'true' values? Or are they the results of measurements, which necessarily contain error? In Clause E.5.2 we read of the 'input quantity w_i' being related to 'its "true" value μ_i' by $w_i = \mu_i + \epsilon_i$, where ϵ_i is an error. So here the word 'input quantity' is being used to describe a datum, not an unknown quantity approximated by this datum.[3] If the input quantity is a datum, say w_i, then the idea in Type B evaluation that probability describes 'degree of belief' about the input quantity refers to doubt about a potential measurement result, and this idea is consistent with the frequentist principles upheld in this book. On the other hand, if the input quantity is the unknown value that is estimated by w_i then the same idea refers to doubt about the value of a fixed but unknown quantity, which is consistent with the distributed-measurand concept. So the way in which the *Guide* was intended to be interpreted and the validity of how it is interpreted in Supplement 1 (JCGM, 2008b) depend strongly on what the words 'input quantity' are to mean. This is a question that must be addressed in any revision of the *Guide*.

In my opinion, another factor that has contributed to the general confusion is the replacement of the idea of 'error' by the idea of 'uncertainty' (see Rabinovich (2007)). An error can either be positive or negative, but this is not true for a component of uncertainty. An error has direction, so when we speak of the standard deviation of an error we picture the spread of measurement results about a fixed value θ. In contrast, the idea of visualizing a component of uncertainty is vague. The lack of orientation or direction that results when we adopt the word 'uncertainty' leads to a temptation to construct a probability distribution for the unknown target value θ centred on a known measurement result x. That is a distributed-measurand idea.

14.3 Consistency with other fields of science

It is instructive to examine the way in which the *Guide* has been perceived in the scientific community, especially by those who have prepared guidelines about uncertainty for organizations within their own fields. Wherever the *Guide*

---

[2] This lack of distinction seems tied up with the deprecation of the concept of an error existing between a quantity and its estimate.

[3] Clause E.5.3 in Appendix E of the *Guide* states that the variances and standard deviations of w_i and ϵ_i are identical, in which case μ_i is not being seen as a variate. The writer of this part of the *Guide* is therefore writing in a way that is entirely in keeping with a frequentist analysis, not with a distributed-measurand analysis.

is referred to outside the metrology community, the understanding seems to accord with the frequentist view of probability. For example, in the preface to the second edition of their book *Experimentation and Uncertainty Analysis for Engineers*, Coleman and Steele (1999, p. 14) write:

The methodology (but not the complete terminology) of the ISO *Guide* has been adopted in standards issued by the American Society of Mechanical Engineers ... and the American Institute of Aeronautics and Astronautics ... and is that presented in this book.

They then describe a treatment of systematic uncertainty that involves the allied concepts of distributed estimates and distributed errors, but not the concept of a distributed measurand. They write (p. 39)

A useful approach to estimating the magnitude of a systematic error is to assume that the systematic error for a given case is a single realization drawn from some statistical parent distribution of possible systematic errors, ...
   In this book we assume that the distribution of possible systematic errors is Gaussian. We are essentially assuming that the systematic error for a given error source is a fixed, single realization from a random distribution of possible errors.

So Coleman and Steele see an approach based on error as being in keeping with the *Guide*. Similar comments are applicable to the book of Dunn (2005) and that of Dieck (2007), who states in his preface that all the material in his book 'is in full harmony with' the *Guide*. In essence, Coleman and Steele are advocating the approach described in the present book, but here we permit parent distributions of systematic errors to be non-Gaussian.

International administrative bodies also appear to favour the frequentist interpretation. Guidelines on uncertainty management in greenhouse gas inventories issued by the Intergovernmental Panel on Climate Change contain an annex entitled *Conceptual Basis for Uncertainty Analysis*. It is clear that the non-normal distributions advocated there are intended to describe data, not to describe any fixed quantity being estimated using the data (IPCC, 2000, A1.2.2, A1.2.4, A1.2.5). And the *Guide* seems to have been mentioned to support the approach taken (IPCC, 2000, A1.2.4).

14.4 Righting the *Guide*

The Working Group of 1980 was convened to arrive at 'a uniform and generally acceptable procedure for the specification of uncertainty' (BIPM *et al.*, 1995, Foreword). Indeed, the *Guide* does seem to have achieved a degree of standardization for basic measurement problems, but its lack of obvious theoretical basis means that the way in which its procedure might be extended for problems of a more complex nature is unclear. So full standardization remains an elusive goal.

The limitations of the *Guide* must be acknowledged, and any revision of it – which is becoming the subject of some discussion (Rabinovich, 2007; Bich, 2008) and which has been announced (JCGM, 2012) – must address the concerns that have been raised by scientists from both sides of the statistical debate. To this end, any revision of the *Guide* might profitably emphasize that it seeks to use the word 'probability' scientifically and that such usage of the word differs from much of its usage in every-day speech (which often does not stand up to scrutiny). For example the statement 'there is probability 0.8 that the measurement result will exceed the target value' does not admit in science the subsequent (fiducial-type) statement that 'there is probability 0.8 that the target value is less than the result we have obtained'. Unless a scientific attempt is made to justify it, as in fiducial inference or in Bayesian inference with a 'flat' prior distribution, the second usage of the term 'probability' remains colloquial and is a source of confusion when used in a scientific context.

The divisions between those who hold different views about probability are deep (and divisions amongst metrologists about 'true values' seem significant also). Perhaps these divisions are so deep that it is unrealistic to expect a single philosophy to the evaluation of measurement uncertainty to emerge in such an important document as the *Guide*. If so, there would be value in developing a methodology that, in most problems, leads to the same result whatever point of view is held. This methodology could then be justified in parallel internally consistent descriptions, one frequentist and one Bayesian.[4] The current situation – where the basis of the methodology is not made clear and where concepts with incompatible philosophical foundations seem combined into a single procedure – is scientifically unacceptable.

By now, the reader of this book will be aware that I favour the use of frequentist, or classical, statistical methods. The most important step in the revision of the *Guide* for those who take the frequentist view would be clear acknowledgement that the quantities given variances in Type B evaluation of uncertainty are the variates for the associated errors. Only then will Type B evaluation be carried out in the way envisaged by the Working Group and in a way that is compatible with Type A evaluation. A second important step would be acknowledgement that the level of assurance attributed to an interval of uncertainty describes a success rate

---

[4] Some issues regarding standardization seem likely to remain, however. For example adopting the Bayesian approach, as in Supplement 1 to the *Guide* (2008b, clause 6.4.9.4), means that the standard uncertainty in Type A evaluation from a set of n observations with sample variance s^2 is $s/\sqrt{n} \times \sqrt{\{(n-1)/(n-3)\}}$, not $s/\sqrt{n}$ as in the *Guide* itself.

in measurement: a client would then be able to give an interval of uncertainty a practical interpretation.

In brief, the *Guide* can be brought into line with the intention behind Recommendation INC-1, with the general interpretation of the current *Guide* in academia and industry, with the basic philosophy of frequentist statistics, and with the philosophy of this book, by clearly stating that Type B evaluation of uncertainty involves attributing 'degree of belief' (or 'state of knowledge') distributions to errors in repetition of the measurement process, not to unknown constants in the particular measurement at hand. It is therefore disappointing to read that the revision of the *Guide* is to be carried out along the lines of Bayesian statistics (JCGM, 2012). Scientists with reservations about this development should make their concerns known.

I have suggested that a revision of the *Guide* might include parallel frequentist and Bayesian justifications of a single favoured computational procedure for the evaluation of measurement uncertainty. Whether or not this step is taken, any procedure advocated in a revision of the *Guide* should be described in terms that are unambiguous. Particular attention seems required when using the words 'value', 'observation', 'indication' and 'quantity'.

Rabinovich (2007) has raised an issue that relates to the meaning of the word 'value'. He argues that the terms 'true value' and 'value of the measurand' must be distinguished. This issue seems related to the question discussed in Sections 1.1 and 12.4 of whether measurement – which involves the attribution of a value (or values) to the measurand – is the approximation of some existing true thing or is the creation of something. To those who believe in true values, the phrase 'the value of the measurand' will naturally mean 'the true value that remains unknown', but to those who do not believe in true values, the same phrase will presumably mean 'the known value that was attributed to the measurand'.

With regard to the words 'observation' and 'indication', consider a statement such as 'If a quantity is repeatedly sampled during the experiment, so that a set of indications is available ...' (Bich, 2008). This phrase seems to suggest that the indications are manifestations of the quantity and are unrelated to any error process. That brings us back to the meaning of the word 'quantity', which we have already commented on in Section 14.2. The quantity referred to in the statement just quoted is an example of what the *Guide* calls an 'input quantity'. Are the input quantities in the *Guide*'s treatment fluctuating, so that they can be attributed probability distributions to represent frequency distributions, or are they fixed values for which a probability distribution merely represents a pattern of belief?

15

Measurement near a limit – an insoluble problem?

Books like this are only necessary because of the difficulties and disagreements that have beset this field of science. These issues relate to the idea of quantifying uncertainty in a way that is meaningful to the user of a measurement result. This final chapter discusses a measurement situation that brings to the fore some of the philosophical and logical questions lying at the heart of the controversy. I trust that the issues raised provide a fitting conclusion to the book.

Suppose that the target value θ is subject to a constraint and that the difference between θ and the constraining value is comparable to or smaller than the standard deviation of the error distribution. In this situation an interval calculated by the methods described in Chapter 7 might extend into the region beyond the constraint, which will be a region that is not physically meaningful. This possibility seems to have caused many to question the meaning and worth of a realized confidence interval obtained in such an event.

In this chapter, we discuss this possibility in the archetypal case of normal measurement error with known variance. A solution that would be deemed satisfactory by a strict frequentist statistician is described, and a criticism of this solution is used to consider some of the underlying issues. The usual objective Bayesian solution to this problem is then discussed. Subsequently, we consider the possible reasons for making a measurement in such a situation and the types of uncertainty statement that might be meaningful. As might be inferred from the title of this chapter, the author doubts the existence of a general solution that is entirely satisfactory.

15.1 Formulation

In some measurements the target value θ is known to lie in a certain range; e.g. in a measurement of a relative concentration we know that θ must lie in the interval

[0, 1]. The extent of this feasible region will be large compared with the measurement error, so without loss of generality we can consider the situation where there is a single constraint, and we can take this to be the constraint $\theta \geq 0$. Let the measurement error variate $X - \theta$ have mean 0 and standard deviation σ. When $\theta \ggg \sigma$ it is possible that X will take a value sufficiently small for one or both of the limits of the raw realized confidence interval to be negative, in which case the interval will then be truncated. So not every interval that might be obtained would subsequently be regarded as equally reliable. In this event, there is information outside both the data and the model that might reduce our level of assurance in the interval obtained. As a result, Claim 1.3 – on which we have based our analysis so far – will be inapplicable.

To the practically minded person, the rare situation where both limits are negative presents less of a problem than the situation where only the lower limit is negative. When the entire interval falls below zero, we have obviously stumbled upon one of the 5% of occasions on which it was accepted that our method would fail. Consequently, we are at liberty to disregard the interval and to choose another means of analysis without lowering the success rate. But this freedom does not exist when the upper limit is positive. For example what should we do when there is normal error with known σ and the estimate obtained is $x = -1.4\sigma$, so that the raw realized 95% confidence interval for θ is $[-3.36\sigma, \ 0.56\sigma]$ and the interval quoted would be $[0, \ 0.56\sigma]$? We cannot dismiss this final interval as definitely not including θ, but neither does it seem to me that we can be as sure of success as we would have been if both limits had been positive.

This question is best expressed using Definitions 3.1 and 3.2. If there were no constraint on θ then the consistent gamblers of Definitions 3.1 and 3.2 would bet on the interval $[-3.36\sigma, \ 0.56\sigma]$ as if they were 95% sure of its success: they would pay up to 95 pence in order to receive 1 pound if the interval contained θ. However, it is known that $\theta \geq 0$ and, arguably, the gamblers will not be prepared to pay so much. So how much will they pay? In our context of measurement uncertainty, a slightly different question is more important: after obtaining the result $x = -1.4\sigma$, what would be an interval for which the gamblers, or you, would pay up to 95 pence in order to receive 1 pound if the interval were successful?

In what follows, we discuss different approaches to the calculation of an interval of 95% measurement uncertainty in this problem when there is normally distributed error with known variance σ^2. Regrettably, the questions that have just been raised will be left unanswered. This is because no acceptable solution seems to exist, chiefly because of the subjective nature of the betting on which Definitions 3.1 and 3.2 rest. As a consequence, the same troublesome conclusion will apply in more general problems of this type.

15.2 The Feldman–Cousins solution

When a measurement result x is obtained in a process that incurs normal error with known mean zero and known standard deviation σ, the realized two-sided 95% confidence interval usually quoted is the familiar interval

$$[x - 1.96\sigma, x + 1.96\sigma]. \tag{15.1}$$

When θ is known to be non-negative, the interval that is quoted will instead be $[\max\{0, x - 1.96\sigma\}, x + 1.96\sigma]$, which is formed from (15.1) by a natural application of the constraint. If x is less than -1.96σ then this new interval is empty. The incorrect idea that a confidence-interval procedure ignores such a constraint and the fact that in rare cases the interval generated is empty have led some scientists to favour Bayesian solutions to such problems.

Following the presentation of such solutions within the particle-physics community, Feldman and Cousins (1998) gave an exact confidence-interval procedure that can never generate an empty interval. Their method is based on the idea that finding a 95% confidence interval for θ is equivalent to inverting a set of 0.05-level tests of the hypothesis $H : \theta = \theta_c$ for different values of θ_c. The extreme values of θ_c for which H would not be rejected form the limits of a realized 95% confidence interval for θ. The usual interval (15.1) is obtained when the tests are equal-tailed at all values of θ_c, but the Feldman–Cousins interval is obtained when the test becomes progressively one-sided in a certain way as θ_c approaches the bound, until at $\theta_c = 0$ it is fully one-sided. This construction guarantees that the interval is never empty. We now describe this solution in analytical terms, and so avoid the need for interpolation in the tables provided by Feldman and Cousins.

It is convenient to consider the case where $\sigma = 1$. Subsequent adjustment for a different value of σ is simply a matter of applying a scale factor. With $\sigma = 1$, the non-rejection region in a test of H is an interval $[x_1(\theta_c), x_2(\theta_c)]$ satisfying

$$\Phi\{x_2(\theta_c) - \theta_c\} - \Phi\{x_1(\theta_c) - \theta_c\} = 0.95, \tag{15.2}$$

where $\Phi\{\cdot\}$ is the standard normal distribution function. The functions $x_1(\theta_c)$ and $x_2(\theta_c)$ form the upper and lower edges of a 'confidence belt' for θ. Figure 15.1 displays two belts corresponding to different rules for choosing the non-rejection region, one with edges given by the pair of solid lines and the other with edges given by the dashed lines. These belts coincide for $\theta_c > 1.96$. Figure 15.1 also indicates the non-rejection region for H when $\theta_c = 2.5$. For any observation x, the limits of the realized 95% confidence interval for θ are the points on the edges of the belt on the vertical line drawn through x. For example Figure 15.1 shows the interval at $x = -1.5$ associated with the belt with dashed edges.

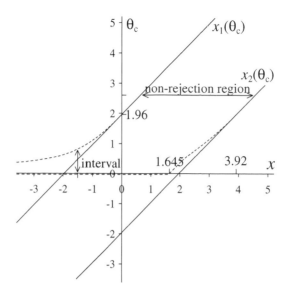

Figure 15.1 Confidence belts for the usual procedure of (15.1) (solid lines) and the Feldman–Cousins procedure of (15.3) and (15.4) (dashed lines).

Different rules for including points in the non-rejection region give rise to different belts and hence to different confidence intervals. In the rule corresponding to (15.1), points are included in order of the value of the probability density function $\phi(z - \theta_c) = (2\pi)^{-1/2} \exp\{-\frac{1}{2}(z - \theta_c)^2\}$, where z is a dummy variable for the observation x. The result is that $x_1(\theta_c)$ and $x_2(\theta_c)$ satisfy $\phi(x_1 - \theta_c) = \phi(x_2 - \theta_c)$, so that x_1 and x_2 are equidistant from θ_c. The confidence belt is then given by the solid lines in Figure 15.1. In the approach of Feldman and Cousins (1998, Section IV: B), the points are included in the non-rejection region according to the value of the ratio of the likelihood when $\theta = \theta_c$ to the maximum likelihood with θ restricted to the feasible region. This ratio is equivalent to the quantity

$$R(z; \theta_c) = \begin{cases} \exp\left\{-\frac{1}{2}(z - \theta_c)^2\right\}, & z \geq 0, \\ \exp\left(z\theta_c - \frac{1}{2}\theta_c^2\right), & z < 0. \end{cases}$$

The non-rejection region $[x_1, x_2]$ for given θ_c then satisfies $R(x_1; \theta_c) = R(x_2; \theta_c)$ instead of $\phi(x_1 - \theta_c) = \phi(x_2 - \theta_c)$, and the corresponding confidence belt is given by the dashed lines in Figure 15.1. Straightforward analysis given in Appendix H shows that the realized confidence interval obtained is $[a, b]$, where

$$a = \begin{cases} 0, & x \leq 1.645, \\ q : \Phi(x - q) - \Phi\left(x - \frac{x^2}{2q} - q\right) = 0.95, & 1.645 < x < 3.92, \\ x - 1.645, & x \geq 3.92 \end{cases}$$
(15.3)

and

$$b = \begin{cases} q: \ \Phi\left(\sqrt{q^2 - 2xq}\right) - \Phi(x - q) = 0.95, & x < 0, \\ x + 1.96, & x > 0. \end{cases} \quad (15.4)$$

The figure 1.645 is the 0.95 quantile of the standard normal distribution, and the figure 3.92 is twice the 0.975 quantile, 1.96. The upper limit b is always greater than 0, so the interval is never empty.

In this way, Feldman and Cousins presented a method that satisfies the strict frequentist requirement of 95% probability of success whatever the value of θ but which never generates an interval with a raw negative limit. So this is a method in which we can have 95% assurance before the measurement. But can we always have 95% postmeasurement assurance in the interval obtained?

Criticism and response

Feldman and Cousins (1998, abstract) note that their method 'unifies the treatment of upper confidence limits for null results and two-sided confidence intervals for non-null results' and called it the 'unified' method. The presentation of the unified method for the Poisson and Gaussian problems of this type led to the publication of other solutions. Mandelkern (2002) provides a summary of other intervals that correspond to different constructions of the non-rejection region.

With regard to the Feldman–Cousins interval in the Gaussian problem, Mandelkern (2002, Sections 2.2, 2.3) writes that for $x < 0$ 'the interval is short, and like that of the Neyman construction, underestimates the uncertainty in ... [θ]' and that the method 'produces limits that are overly restrictive for negative observations'. Similarly, when referring to that method, Woodroofe and Zhang (2002) write 'We agree with Mandelkern, however, that it can produce unbelievably short intervals.' But what criteria exist by which any alternative interval could be deemed to be believable or realistic or to be a proper estimate of 'the uncertainty in θ'? And there seems to be no basis for considering any interval to be unbelievable if the only available information about θ is utilized fully. For such comments to have meaning, external information of some form must be available.

So what makes one numerical limit 'realistic' and another 'unrealistic'? This question is related to the question 'what is actually meant by quantifiable experimental uncertainty?' raised in part by Wasserman (2002) in his comment on the article by Mandelkern. Experimentalists envisage providing a statement of 'experimental uncertainty' with an attached level of assurance, say 95%. But what is such a statement actually intended to mean? I asked this question in Section 1.6, and many statisticians would ask the same question. An answer that is useful in most measurement situations – where there is no information about θ except that found in the model and data – was given in Claim 1.3 and emphasized in point 11 of the list of principles in Section 6.1.

A modification for greater realism?

The interval $[a, b]$ is obtained by including points in the non-rejection region in order of the value of a certain likelihood ratio, $R(z; \theta_c)$. However, we are at liberty to choose a different way of forming the non-rejection region. In particular, we might fix one edge of the belt and then find the other edge by solving (15.2) for each point. The first edge might be fixed in the following way.

Suppose we were to accept the criticism that the upper limit b is sometimes too small to be realistic or plausible. That would require the existence of some principle by which we could make such a judgement. Presumably, on slowly raising this limit we eventually obtain a value h deemed realistic according to that principle. The function $h(x)$ for all x could then be constructed in a thought experiment, and this function could form the upper edge of the confidence belt. The other edge of the belt would then be found by solving (15.2). So, if there is any objective principle whereby the upper limit can be deemed realistic, the basic approach adopted by Feldman and Cousins can be applied to give an acceptable solution. And if there is no such principle then any criticism that certain intervals are 'unrealistic' does not carry a great deal of weight.

But what objective principle exists for quantifying the level of postmeasurement assurance that we can have in a problematic realized confidence interval? And, more importantly, what principle is there for adjusting the interval so that it becomes 95% reliable in the sense of Definitions 3.1 and 3.2? As explained in Chapter 13, the idea that postmeasurement assurance can be quantified objectively by the proper use of information is the basic claim of 'objective' Bayesian statistics. So we now discuss the standard objective Bayesian approach to this problem.

15.3 The objective Bayesian solution

The Bayesian statistician must specify a prior density function to represent belief about the target value θ before the datum x was obtained. In an objective Bayesian analysis this is likely to be the improper density function

$$f_{\text{prior},\theta}(z) = \begin{cases} \text{constant}, & z \geq 0, \\ 0, & z < 0. \end{cases} \quad (15.5)$$

In using this prior density function, the relevant person is saying 'Prior to the experiment, I considered every non-negative value to be equally likely for θ.' The

15.3 The objective Bayesian solution

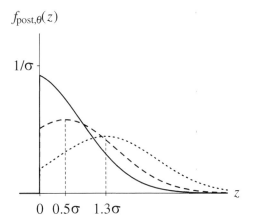

Figure 15.2 Posterior probability density function of θ. Solid line: $x = -0.2$. Dashed line: $x = 0.5\sigma$; dotted line $x = 1.3\sigma$.

resulting posterior distribution of θ is the normal distribution with mean x and variance σ^2 truncated below at 0 and renormalized. The posterior density function of θ is therefore

$$f_{\text{post},\theta}(z) = \begin{cases} \dfrac{1}{\sigma} \cdot \dfrac{\phi\{(z-x)/\sigma\}}{\Phi(x/\sigma)}, & z \geq 0, \\ 0, & z < 0, \end{cases} \qquad (15.6)$$

which is shown for three different values of x/σ in Figure 15.2. Subsequently a 95% credible interval for θ will be calculated according to some formal or informal rule. One natural rule is to always choose the interval containing the central 95% of the distribution. Another is to always choose the shortest interval.

Criticism

Criticism of this approach is likely to focus on the choice and meaning of the flat prior distribution (15.5). A number of points seem relevant.

1. The imposition of the constraint $\theta \geq 0$ removes the arbitrary nature of the position of the origin from the problem of estimating a location, and so it removes a major factor in the argument for a flat distribution that might be applicable in the unconstrained case.
2. In many situations the object measured will have deliberately been constructed to make the value of the measurand close to zero. (For example a filter that removes light in one range of frequencies will be constructed to produce negligible attenuation elsewhere.) In such a situation a smaller value for θ must

be deemed to be more likely than a larger value. The attribution of a flat prior distribution to θ ignores such an intention. In effect, it treats the value of the measurand as being the outcome of a random non-negative number generator!

3. In other situations the measurand will be the level of an undesirable perturbing phenomenon. (For example the measurand might be the concentration of a toxin in a water sample.) It seems unduly pessimistic to consider the existence of every possible level of this phenomenon to be equally probable.

4. More generally, we can consider the value of the measurand to have been created by some process, natural or otherwise, that operates until the value settles at θ. If the flat distribution is 'correct' then the effect of the barrier at $z = 0$ on this process must be to reflect it back along the positive real axis. If the barrier absorbs instead of reflects then the prior density function should have a spike (delta function) of some size at $z = 0$, while if the barrier exerts a retarding influence on this process as z approaches zero then the prior density function should be decreasing from $z = 0$. (For example suppose that an attempt has been made to create a vacuum in a cylinder and that the measurand is the residual air pressure. The process of removing air particles gets progressively more difficult, so this process seems to have a greater probability of finishing when the pressure lies in the interval $[a, a + \delta)$ than in the interval $[2a, 2a + \delta)$.)

5. As formulated, the problem involves the constraint $\theta \geq 0$, not the constraint $\theta > 0$. The use of the flat prior distribution attributes no probability to the event $\theta = 0$. So the use of this distribution seems more suited to the second situation than the first.

6. And might the target value be exactly zero? If so then the prior distribution must have a spike at $\theta = 0$. If the prior distribution has no spike, as in (15.5), then the posterior distribution cannot have a spike either, and so the posterior probability that $\theta = 0$ will be zero. Whatever the value x, it will then be deemed effectively certain that $\theta > 0$, which is inappropriate.

So we conclude that there are a number of types of measurement for which a density function that is decreasing or that has a spike at the origin would be more appropriate. This implies that the flat distribution is not faithfully representing the prior information available. So why should the posterior distribution (15.6) be faithfully representing the information available after observation of the datum x? And, to echo the basic criticism of objective Bayesian statistics given in Chapter 13, we must ask what evidence exists that an objective Bayesian analyst would be prepared to act on the corresponding posterior probabilities in the manner implied in Definitions 3.1 and 3.2.

15.4 Purpose and realism

So far we have not questioned the need to calculate a 95% interval of measurement uncertainty. Neither have we raised the possibility that a semi-infinite interval like $[x - 2.33\,\sigma, \infty)$ might be more relevant than a finite interval. Nevertheless, there are many practical measurements where the goal requires something other than a finite interval, and it may be that some of the philosophical obstacles we are encountering can be negotiated by focusing on the purpose of the measurement.

Testing

The purpose of the measurement might be to compare the value of the measurand with a single predetermined small level having some physical significance, say the level a. So the real question of interest might be 'How much evidence is there that $\theta > a$?', as in the task of determining whether neutrinos are massless (Mandelkern, 2002), where $a = 0$. If this is the case then the task appears to be one of detection, not measurement. The idea of 'detection' implies that the objective is to disprove the null hypothesis that no stimulus is present, and this approach is biased towards the null point of view. However, the idea of 'measurement' seems unbiased in this respect.

So how useful is the concept of an interval of measurement uncertainty in testing whether $\theta > a$? The task is not one of finding a finite interval admitting a fixed level of assurance, which is the basic concept of measurement uncertainty discussed in this book. Rather the task is one of testing the notion that $\theta > a$, which means assessing our level of assurance that the target value lies in the interval $[a, \infty)$. So – as was exemplified in Section 9.4, where we considered conformity assessment – we see that 'testing' is an area where the standard view of uncertainty in measurement does not fit the final purpose of the measurement.

One-sided intervals

A different kind of question is 'what is the value z for which we have 95% assurance in the assertion "$\theta > z$"?' Here we are seeking to quote a one-sided interval $[z, \infty)$ as a 95% interval of measurement uncertainty. The interval required in a decision-making problem is often of this form (see Fraser *et al.* (2004)). Such an interval can be quoted from a two-sided 90% interval constructed in a probabilistically symmetric way: if $[a, b]$ is a 'symmetric' 90% interval then we can state that 'I am 95% sure that $\theta \geq a$' and 'I am 95% sure that $\theta \leq b$', so that $[a, \infty)$ and $[0, b]$ are 95% one-sided intervals for θ. However, no such statement could be made if we had adopted the approach of Feldman and Cousins in the generation of a 90% interval or if we had adopted the Bayesian approach and quoted a 90% credible interval different from the interval containing the central 90% of the distribution.

Combination of measurement results

Is a summary statement of measurement uncertainty always necessary? When we measure an important quantity and our measurement is one of several measurements made by various groups, the overall task will be to combine the results to form a best current estimate and to associate an uncertainty statement with that combined estimate (e.g. Mandelkern (2002)). The proper way to record the output of our measurement will be to state our raw result (even if it is negative) along with our data and the associated error model. There will be no need to associate an interval of measurement uncertainty with our individual result, nor would any such interval be helpful unless we could recover all that was relevant about our measurement from it.

That idea seems to warrant restatement and emphasis. There is no meaning in a statement of uncertainty about a target value if that statement ignores all other acceptable estimates of this target value. (Of course, this does not apply in the context of a measurement comparison.) A reader who protests might be understanding 'uncertainty' in the way described in Section 12.4.

Model inadequacy

Mandelkern (2002) observes that physicists are cautious about biasing an analysis by changing the model after observing the data. This caution may be contributing significantly to the existence of problematic intervals, which are intervals that will arise only infrequently if the models are adequate. Van Dyk (2002), in his discussion of Mandelkern's paper, wonders how frequently major physics experiments result in unsatisfactory raw intervals and points out that an empty realized confidence interval is strong evidence that the model is inadequate. To van Dyk, the creation of new estimation procedures for use with a model whose accuracy is severely questioned by the data seems misguided. In effect, he is urging the physicist to be realistic in matters of data analysis.

Repeating the measurement

Suppose we have obtained an interval that is problematic because of the existence of a known bound on θ. A pragmatist might advocate repeating the experiment and combining the results until the interval calculated seems acceptable. The value of this approach is open to debate, and once again the purpose of the measurement will be an important consideration. This approach would introduce some bias into the overall estimation procedure but reduce the variance of this procedure. It would also violate a requirement of most frequentist techniques that the sample size be fixed in advance, the result being that the actual confidence level of the procedure might differ from the stated figure of 95%. (The idea of terminating an experiment when the results appear favourable is a form of cherry picking.) Furthermore, the repetition of the experiment will not solve the underlying problem if fixed errors are dominating.

Summary

So there are several reasons why the existence of a raw interval with an unacceptable limit can be less troublesome than first thought. The difficulty that we have been discussing might be avoided by considering the purpose of the measurement, which might call for an expression of measurement uncertainty that differs from a finite interval with a fixed level of assurance. Moreover, the existence of an improbable observation can relate to inadequacy in the model. If this is the case then the model should be reconstructed.

15.5 Doubts about uncertainty

The statements made by Mandelkern (2002) on behalf of his physics colleagues, the questions that he raises and the responses of statisticians who commented on his article (e.g. Gleser (2002), Van Dyk (2002), Wasserman (2002), Woodroofe and Zhang (2002)), all indicate that the quantification of uncertainty in science is problematic. We must ask what the word 'uncertainty' – which in English is an abstract uncountable noun – means when it is used in a way that suggests its meaning in concrete and quantitative.

So what is experimental uncertainty?

Mandelkern (2002, p. 151) writes that physicists would like an interval quoted to 'convey an estimate of the experimental uncertainty' and also that in a frequentist analysis 'the variance of the parent distribution ... reflects the experimental uncertainty'. But we have seen that the variance of the parent distribution does not necessarily reflect postmeasurement uncertainty when there is a bound on the target value. So is the term 'experimental uncertainty' implicitly being given two different meanings? In the context of making a globally-best estimate of some quantity (e.g. a constant of nature) Mandelkern suggests that the uncertainties of all contributing results are needed. This seems to require associating the concept of uncertainty exclusively with the spread of experimental results, and this implies that the presence of the bound is of no consequence. If the presence of the bound is to be taken into account then the concept of uncertainty becomes something else, namely, the concept of postmeasurement doubt about the target value.

Thus, there seems to be doubt as to what 'experimental uncertainty' actually is. Indeed, Wasserman (2002) responded to Mandelkern by considering what this term might mean. Scientific language may need to be clarified in order to make progress. Are we to equate 'experimental uncertainty' with 'experimental variability' (as broadly in the frequentist view) or should other information also be taken into account? And can we equate 'experimental uncertainty' with 'measurement uncertainty' in our context?

I believe that this problem is exacerbated by the choice of the word 'uncertainty'. Nothing seems more subjective than the notion of uncertainty, yet many measurement scientists give the impression that, for them, uncertainty is something objective, something that can be determined uniquely by all reasonable observers and subsequently used as if it were something real. Do these ideas withstand scrutiny?

Is there a unique uncertainty?

There is, perhaps, a tendency to think that *the uncertainty of measurement* is a unique thing that is able to be defined and calculated. This idea should be examined. According to some definition of 'best', the measurement result will be our unique best estimate of the target value θ. But can we say something analogous about the calculated figure of uncertainty? Are we claiming that this figure of uncertainty is unique and that it is some optimal estimate of something real? As now argued, the answer depends on whether it is the uncertainty associated with any particular *measurement*, which has been the context assumed in most of this book, or whether it is the uncertainty associated with any particular *measurement technique*.

When estimating a quantity θ from a small set of experimental data taken to have an approximately normal parent distribution, as in the typical Type A evaluation of uncertainty, the length of the realized confidence interval for θ depends on the spread in the data. The interval will be short on some happy occasions and long on other occasions. This fact is conclusive evidence that 'the measurement uncertainty' – as reflected in the length of a realized confidence interval – is not simply a property of the measurement problem or the measurement technique. The length of the interval is, in part, a matter of chance: it is simply the result of a procedure that had a random element. So a figure of uncertainty based on statistical data that is obtained in the measurement at hand does not represent anything concrete or real. However, the same cannot be said for a figure of uncertainty that is to be routinely associated with an established measurement technique: when the technique is unbiased this figure will be understood to be some multiple of the parent standard deviation of the error. Once again, we see the existence of different concepts of 'uncertainty'.[1]

Uncertainty and error

Thus, there are different possible concepts of experimental uncertainty and there are different possible contexts for figures of standard uncertainty in measurement. Is it possible that the word 'uncertainty' – being properly abstract – is seen as a suitable catch-all word? This word seems to have usurped the place of a word with a much clearer meaning, 'error' (Rabinovich, 2007). Once we have conceived of a target value, the term 'error' is well defined and the error in any practical measurement is something unique and real. Arguably, using the term 'uncertainty' in its stead has promoted a great deal of confusion: the word 'uncertainty' seems better left with its colloquial abstract meaning.

15.6 Conclusion

We have reached the end of this book. Once more, it seems appropriate to state the basic question of Section 1.6, which is related to all of our analysis: *'what is to be understood when a specified level of assurance is attached to an interval of uncertainty provided with a measurement estimate?'* In effect, the question is answered by suggesting that the person using the uncertainty interval would behave in the manner implied in Definitions 3.1 and 3.2 – and by emphasizing that a certain long-run success rate is implied.

But this conclusion presents a problem when there is information about the value of the measurand found outside the experimental data and the statistical model, as in the situation addressed in this chapter, where there is a known bound on the value of the measurand. In this situation, our level of assurance *before* calculation that the interval *will* contain θ, which is the frequentist concept of confidence, does not always seem applicable to the numerical interval subsequently obtained. And the problems of representing the genuine prior belief of the person who will use the result means that the output of a Bayesian analysis lacks meaning. So, for

---

[1] One of the questions asked by Mandelkern (2002, p. 158) relates to this question of whether a figure of measurement uncertainty is unique. He asks whether a favourably improbable observation gives greater knowledge of an unknown parameter than a more typical observation. The answer must be 'yes', because – as just seen in our text – the idea is well accepted that randomness plays a legitimate role in determining the length of a realized confidence interval in the Type A case. Van Dyk (2002) also answers in the affirmative, but he gives a different justification.

me, no method presented in this chapter is fully satisfying (though the Feldman–Cousins method comes close). In fact, I do not think that a procedure meeting our requirements can exist in this measurement situation. Postmeasurement assurance only seems definable in terms of the theoretical human behaviour described in Definitions 3.1 and 3.2, but no practical approach seems to generate intervals for which the user of the measurement result would be prepared to behave in the way implied.

Therefore, I have concerns about the evaluation of 'uncertainty' when making measurements close to a physical limit. Having just raised these concerns, I now ask the reader to put them to one side! In the more typical measurement problem, any limit on the value of the measurand will, relative to the standard deviation of the error distribution, be so distant from the measurement result that this issue can be ignored. There will be no relevant information about the value of the measurand outside the data and model, so our postmeasurement level of assurance will be equal to our premeasurement level of assurance. The classical concept of 'confidence' – when properly extended for the realities of measurement – will be adequate. This provides a justification for the approach taken in this book, which has described principles and methods developed under an 'extended classical' point of view.

Appendix A
The weak law of large numbers

In Section 3.1 it was stated that the average profit of a person engaging in a long series of independent contracts converges to the average of the expected profits for those contracts, whether those contracts are identical or not. This is a key claim, for our claim about the meaning of a probability rests upon it. This claim can be shown to be correct using the *weak law of large numbers*, which can be stated as follows (see Wilks (1962), p. 99).

Let $\{Y_1, Y_2, \ldots\}$ be a sequence of independent variates in which Y_i has mean 0 and variance σ_i^2, and define $\bar{Y}_n \equiv \sum_{i=1}^{n} Y_i/n$. If $\lim_{n \to \infty} \sum_{i=1}^{n} \sigma_i^2/n^2 = 0$ then the sequence of variates $\{\bar{Y}_1, \bar{Y}_2, \ldots\}$ converges in probability to 0, i.e. $\lim_{n \to \infty} \Pr(|\bar{Y}_n| > \epsilon) = 0$ for any $\epsilon > 0$.

The word 'independent' can be replaced by the word 'uncorrelated' (Wilks, 1962, p. 100). Clearly, the condition that $\lim_{n \to \infty} \sum_{i=1}^{n} \sigma_i^2/n^2 = 0$ is satisfied when each Y_i is bounded between the same two values.

Consider a set of independent random gains $\{X_i\}$ in which X_i has expectation $\mathcal{E}(X_i)$. The law can be applied with $Y_i = X_i - \mathcal{E}(X_i)$. We find that if each X_i is bounded between any single pair of values then, for any $\epsilon > 0$,

$$\lim_{n \to \infty} \Pr\left(\frac{1}{n} \sum_{i=1}^{n} \mathcal{E}(X_i) - \epsilon < \frac{1}{n} \sum_{i=1}^{n} X_i < \frac{1}{n} \sum_{i=1}^{n} \mathcal{E}(X_i) + \epsilon \right) = 1.$$

This shows that there is convergence of the average gain of a person engaging in a set of independent bets or contracts to the average of the expected values for those bets. This holds whether the bets are identical or different, i.e. whether there is one repeated experiment or a sequence of different experiments.

Appendix B
The Sleeping Beauty paradox

The Sleeping Beauty paradox involves the concept of subjective probability. Knowing the rules, Sleeping Beauty agrees to take part in an experiment. She is sent to sleep on Sunday. Subsequently a fair coin is tossed and the outcome recorded. If the coin turns up heads she is woken on Monday, and asked the question, 'how strong is your belief that the coin turned up heads?' She then answers the question and is subsequently sent back to sleep using a drug that erases her memory of Monday and all its events. Whichever side of the coin turns up, she is woken on Tuesday and asked the same question, which she answers. What should her answer or answers be?

Many would say that her answer should be 'one-half'. But if Sleeping Beauty and her questioners see Definition 3.1 as defining strength of belief then the question must be replaceable by the contract: 'We will tell you whether the coin turned up heads. Up to how many pounds will you pay to receive 1 pound if it were heads?' Sleeping Beauty knows that if the coin turns up heads she answers the question twice and if it turns up tails she answers it once. Also she knows that if the experiment were repeated many times the coin would turn up heads on 50% of occasions. So if she pays q pounds then her expected value of profit is $0.5 \times 2 \times (1-q) + 0.5 \times (-q)$, which is positive for $q < 2/3$. So her answer must be 'two-thirds'.

Thus, if Sleeping Beauty wished to maximize her expected profit from the experiment then she would (a) declare her strength of belief to be 2/3 and (b) point out that the only way she can prove that she is not lying is if the gamble is arranged and the experiment is repeated many times. She could then continually wager any amount up to 66 pence and profit from those who persist in the belief that her answer should be 1/2!

This paradox is relevant to the idea that the set of opportunities is important in an assessment of probability. Sleeping Beauty recognizes that the coin would turn up heads in two-thirds of her opportunities.

Appendix C
The sum of normal and uniform variates

Suppose X is the sum of a standard normal variate and an independent continuous uniform variate on the interval $[-h, h]$. Then X is symmetric with mean 0, variance $1 + h^2/3$, and density and distribution functions (Bhattacharjee et al., 1963)

$$f_X(z) = (2h)^{-1} \{\Phi(z+h) - \Phi(z-h)\},$$
$$\Pr(X \leq z) = (2h)^{-1} \{G(z+h) - G(z-h)\}, \qquad G(y) \equiv y\Phi(y) + \phi(y),$$

with $\phi(\cdot)$ and $\Phi(\cdot)$ being the standard normal density and distribution functions. Figure C.1 shows $f_X(z)$ for different values of h.

It follows that if $E_1 \sim N(0, a^2)$ and, independently, $E_2 \sim U(-b, b)$ then $E = E_1 + E_2$ has mean zero, variance $a^2 + b^2/3$, and has

$$f_E(z) = 1/(2b) \{\Phi(z/a + b/a) - \Phi(z/a - b/a)\},$$
$$\Pr(E \leq z) = a/(2b) \{G(z/a + b/a) - G(z/a - b/a)\}.$$

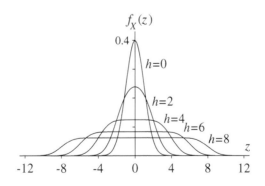

Figure C.1 Probability density function $f_X(z)$ for different values of h.

Appendix D

Analysis with one Type A and one Type B error

Consider the measurement of $\theta = \theta_1 + \theta_2$, where (i) θ_1 is estimated by the mean, x_1, of n independent observations drawn from a normal distribution with mean θ_1 and *unknown* standard deviation σ and (ii) θ_2 is determined by expert knowledge to lie within a known amount w of some estimate x_2, with no value being thought more plausible than any other. So the errors $x_1 - \theta_1$ and $x_2 - \theta_2$ are taken to be drawn from $N(0, \sigma^2/n)$ and $U(-w, w)$.

Set $x = x_1 + x_2$. The total error $x - \theta$ is the realization of a variate $E = (\sigma/\sqrt{n})Z + U$, where $Z \sim N(0, 1)$ and $U \sim U(-w, w)$. Let s be the sample standard deviation associated with x_1 and let S be the corresponding variate. Suppose we can find a function $H(S, w, n)$ satisfying

$$0.95 \leq \Pr\{H(S, w, n) \geq |E|\} \leq 0.95 + \delta \quad \text{(D.1)}$$

for every possible σ and for some δ. The interval with limits $x \pm H(s, w, n)$ will then be the realization of a valid 95% confidence interval for θ.

Choosing $H(S, w, n)$ to minimize δ will encourage narrow intervals. Ideally this function will be equal to $t_{n-1, 0.975} S/\sqrt{n}$ when $\sigma/\sqrt{n} \gg w$ and equal to $0.95w$ when $\sigma/\sqrt{n} \ll w$. Simulations for $5 \leq n \leq 100$ and $0.05\sigma/\sqrt{n} \leq w \leq 256\sigma/\sqrt{n}$ suggest that (D.1) holds with $\delta = 0.004$ when (Willink, 2006a)

$$H(S, w, n) = \sqrt{t^2_{n-1, 0.975} S^2/n + (0.95w)^2 + G(S, w, n)}$$

with $G(S, w, n)$ being the quantity

$$\left(0.37 - \frac{1.295}{n} - \frac{3.7}{n^2}\right) \left(\frac{t_{n-1, 0.975} S}{\sqrt{n}}\right)^{0.5} (0.95w)^{1.5} \exp\left(\frac{-0.35 \times 0.95w}{t_{n-1, 0.975} S/\sqrt{n}}\right).$$

The term $G(S, w, n)$ was chosen empirically for good performance. Presumably, a simpler term would be available if we accepted a larger δ.

A practical system of analysis based on this idea would need to accommodate several Type A errors and a Type B error with a more general form.

Appendix E

Conservatism of treatment of Type A errors

As described in Section 6.4, our treatment of a Type A error-variate E_i associated with normal sampling involves taking E_i to have the distribution of $(s_i/\sqrt{n_i})T_{n_i-1}$. Our conjecture is that this leads to valid 95% intervals. So we wish to show that $|e| \leq q(s_1, \ldots, s_k)$ on at least 95% of occasions for all m, k, $\{\sigma_i\}$, $\{n_i\}$ and for all symmetric Type B error-variates $E_{k+1}, \ldots, E_m$.

The total Type B error-variate, $\sum_{i=k+1}^{m} a_i E_i$, will be symmetric and can be thought of as a single variate. So without loss of generality (except perhaps in the form of distributions used) we can set $m = k + 1$. Simulations were therefore carried out with $k = 2$ and $m = 3$. Without loss of generality we could set $\sigma_1/\sqrt{n_1}$ and each a_i to unity. Each scenario was defined by the parameters n_1, n_2 and $\sigma_2^* \equiv \sigma_2/\sqrt{n_2}$ and by the distribution of the overall Type B error. Four forms were chosen to represent this Type B error: Student's t distribution with five degrees of freedom, the normal distribution, the uniform distribution and the arc-sine distribution. Because a mean of zero is assumed, the distribution is fully specified by its variance σ_3^2.

For each form of distribution and for each parameter vector $(n_1, n_2, \sigma_2^*, \sigma_3)$ with $n_1, n_2 \in \{2, 4, 10, 100\}$ and $\sigma_2^*, \sigma_3 \in \{0.1, 1, 10\}$, the proportion of times that $|\tilde{e}| < \hat{q}(\tilde{s}_1, \ldots, \tilde{s}_k)$ in 10^9 non-independent trials was recorded. Here $\hat{q}(\tilde{s}_1, \ldots, \tilde{s}_k)$ is an estimate of $q(\tilde{s}_1, \ldots, \tilde{s}_k)$ based on ordering a sample of size 9999. Of the $4^3 \times 3^2 = 576$ combinations, 523 gave proportions greater that 0.95 and 116 gave proportions greater than 0.98. Only nine gave proportions less than 0.947. The underlying proportions for these nine combinations were re-estimated using estimates of $q(\tilde{s}_1, \ldots, \tilde{s}_k)$ based on ordering a sample of size 99999. The results ranged from 0.9497 to 0.9523 and had mean 0.9509. The same method was applied in nine situations with $\sigma_2^* = \sigma_3 = 0$, where the underlying proportion is exactly 0.95. The results ranged from 0.9491 to 0.9524 and had mean 0.9504. This is substantial evidence in support of the conjecture.

Appendix F
An alternative to a symmetric beta distribution

The cumulants method presented in Section 7.3 can be criticized for potentially involving a distribution with an unrealistic form. For values of γ between -1.2 and 0, the symmetric Pearson distribution is a unimodal beta distribution, and such a distribution does not necessarily have a form that can eventuate when error variates are added. A solution is to replace this family of beta distributions by a family comprising symmetric distributions that arise as convolutions and containing the uniform and normal distributions as extreme cases. The distributions are to have forms described by one shape parameter, which is the coefficient of excess.

One possibility is the family of distributions obtained when adding a normal variate and an independent uniform variate (e.g. Fotowicz (2006)). (See Appendix C.) A variate with the normal distribution $N(0, a^2)$ has mean 0, variance a^2 and fourth cumulant zero, while a variate with the uniform distribution $U(-b, b)$ has mean 0, variance $b^2/3$ and fourth cumulant $-2b^4/15$. Let E be the

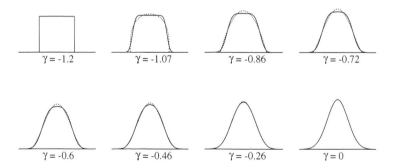

Figure F.1 Probability density functions of E (solid) and the symmetric Pearson variable (dashed) with the same value of γ. All the distributions shown are scaled to have equal variance.

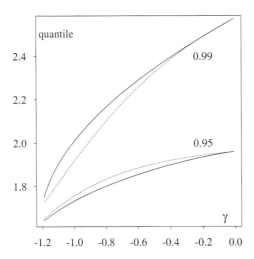

Figure F.2 The 0.975 and 0.995 quantiles of the standardized distribution of E (solid) and the standardized symmetric Pearson distribution with the same value of γ (dotted).

sum of these variates. So E has mean zero, variance $a^2 + b^2/3$ and fourth cumulant $-2b^4/15$. Then E has coefficient of excess

$$\gamma = \frac{-2/15}{(a^2/b^2 + 1/3)^2},$$

which varies smoothly from the uniform value of -1.2 to the normal value of 0 as a/b increases from 0 to ∞. Figure F.1 shows various distributions of E and shows the symmetric distributions from the Pearson system with the same values of γ.

The standardized variate is $E/\sqrt{(a^2 + b^2/3)}$. Figure F.2 shows the 0.975 and 0.995 quantiles of this variate as a function of γ and also shows the 0.975 and 0.995 quantiles of the standardized Pearson approximation given by (7.20) and (7.21).

Appendix G
Dimensions of the ellipsoidal confidence region

The hyperellipsoid defined by equating the left-hand and right-hand sides of (10.1) extends in the ith dimension across the interval

$$\left[\bar{X}_i - \sqrt{T^2_{n-1,p,0.95}} S_i/\sqrt{n},\ \bar{X}_i + \sqrt{T^2_{n-1,p,0.95}} S_i/\sqrt{n}\right],$$

which is (10.2). This result is difficult to find in texts, so we shall give a proof of it here.

Consider the p-dimensional hyperellipsoid formed by the end-points of the column vectors $\mathbf{y} \equiv (y_1, \ldots, y_p)'$ that satisfy $\mathbf{y}'\mathbf{A}^{-1}\mathbf{y} = b$, where $\mathbf{A}^{-1}$ is a known symmetric $p \times p$ matrix and b is a known scalar. The largest value of y_1 on this hyperellipsoid occurs at the single point where

$$\frac{\partial \mathbf{y}'\mathbf{A}^{-1}\mathbf{y}}{\partial y_j} = 0, \qquad j = 2, \ldots, p.$$

So if $\mathbf{u}_j$ is the column vector of length p with jth element equal to 1 and all other elements equal to 0 then at this point $\mathbf{u}'_j\mathbf{A}^{-1}\mathbf{y} + \mathbf{y}'\mathbf{A}^{-1}\mathbf{u}_j = 0$ for $j = 2, \ldots, p$, which in turn means that $\mathbf{u}'_j\mathbf{A}^{-1}\mathbf{y} = 0$ for $j = 2, \ldots, p$ because $\mathbf{A}^{-1}$ is symmetric. Let $\mathbf{Q}$ be the $p \times p$ matrix $[\mathbf{y}, \mathbf{u}_1, \ldots, \mathbf{u}_p]'$, and let $\mathbf{b}$ be the column vector $(b, 0, \ldots, 0)'$ of length p. Then this extremal point $\mathbf{y}$ satisfies the set of p linear equations given by $\mathbf{Q}\mathbf{A}^{-1}\mathbf{y} = \mathbf{b}$. Therefore, $\mathbf{y} = \mathbf{A}\mathbf{Q}^{-1}\mathbf{b}$. Now $\mathbf{Q}^{-1} = [\mathbf{w}, \mathbf{u}_1, \ldots, \mathbf{u}_p]'$, where $\mathbf{w} = (1/y_1, -y_2/y_1, \ldots, -y_p/y_1)'$. Thus $\mathbf{y} = \mathbf{A}\mathbf{b}/y_1$. The first of the p relationships described by this equation is $y_1 = A_{1,1}b/y_1$, where $A_{1,1}$ is the first element on the leading diagonal of $\mathbf{A}$. So $y_1 = \sqrt{(A_{1,1}b)}$, which means that the hyperellipsoid extends in the first dimension from $-\sqrt{(A_{1,1}b)}$ to $\sqrt{(A_{1,1}b)}$. Similarly, it extends in the ith dimension from $-\sqrt{(A_{i,i}b)}$ to $\sqrt{(A_{i,i}b)}$.

We can identify $\mathbf{y} = \bar{\mathbf{X}} - \boldsymbol{\theta}_c$, $\mathbf{A} = \mathbf{S}/n$, $A_{i,i} = S_i^2/n$ and $b = T^2_{n-1,p,0.95}$. This gives the result stated.

Appendix H
Derivation of the Feldman–Cousins interval

The limits of the Feldman–Cousin interval $[a, b]$ are given by (15.3) and (15.4). All but two of the branches of these equations can be derived directly from study of Figure 15.1. The exceptions are the expression for a when $1.645 < x < 3.92$ and the expression for b when $x < 0$.

The lower limit a is the value of θ_c such that $x = x_2(\theta_c)$. So a is the value of θ_c for which $\Phi(x - \theta_c) - \Phi(x_1 - \theta_c) = 0.95$, where x_1 is such that $R(x_1; \theta_c) = R(x; \theta_c)$, i.e., $2 \log R(x_1; \theta_c) = 2 \log R(x; \theta_c)$. Now

$$2 \log R(v; \theta_c) = \begin{cases} -(v - \theta_c)^2, & v \geq 0, \\ 2v\theta_c - \theta_c^2, & v < 0. \end{cases}$$

In the branch of interest $x > 0$ and $x_1 < 0$, from which we find that $x_1 = x - x^2/(2\theta_c)$, which leads to the expression in (15.3).

The upper limit b is the value of θ_c such that $x = x_1(\theta_c)$. So b is the value of θ_c for which $\Phi(x_2 - \theta_c) - \Phi(x - \theta_c) = 0.95$, where x_2 is such that $R(x; \theta_c) = R(x_2; \theta_c)$. In the branch of interest $x < 0$ and $x_2 > 0$, from which we find that $x_2 = \theta_c + (\theta_c^2 - 2x\theta_c)^{1/2}$. This leads to the expression in (15.4).

References

Acton, F. S. 1959. *Analysis of Straight-Line Data*. Wiley.
Alfassi, Z. B., Boger, Z., and Ronen, Y. 2005. *Statistical Treatment of Analytical Data*. Blackwell.
Barnett, V. 1973. *Comparative Statistical Inference*. Wiley.
Bayarri, M. J., and Berger, J. O. 2004. The interplay of Bayesian and frequentist analysis. *Statistical Science*, **19**, 58–80.
Beers, Y. 1957. *Introduction to the Theory of Error*, 2nd edn. Addison Wesley.
Bernardo, J. M. 2005a. An integrated mathematical statistics primer: Objective Bayesian construction, frequentist evaluation. In: *Proceedings of the 55th International Statistical Institute Session, Sydney, 2005*. ISI.
Bernardo, J. M. 2005b. Reference analysis. Pages 17–90 of: Dey, D. K., and Rao, C. R. (eds.), *Handbook of Statistics*, vol. 25: *Bayesian Thinking: Modeling and Computation*. Elsevier.
Bevington, P. R. 1969. *Data Reduction and Error Analysis for the Physical Sciences*. McGraw-Hill.
Bhattacharjee, G. P., Pandit, S. N. N., and Mohan, R. 1963. Dimensional chains involving rectangular and normal error-distributions. *Technometrics*, **5**, 404–406.
Bich, W. 2008. How to revise the GUM. *Accreditation and Quality Assurance*, **13**, 271–275.
BIPM, IEC, IFCC *et al*. 1995. *Guide to the Expression of Uncertainty in Measurement*.
Box, G. E. P., and Draper, N. R. 1987. *Empirical Model-building and Response Surfaces*. Wiley.
Chatfield, C., and Collins, A. J. 1980. *Introduction to Multivariate Analysis*. Chapman and Hall.
Colclough, A. R. 1987. Two theories of experimental error. *Journal of Research of the National Bureau of Standards*, **92**, 167–185.
Coleman, H. W., and Steele, Jr., W. G. 1999. *Experimentation and Uncertainty Analysis for Engineers*, 2nd edn. Wiley.
Croarkin, C. 1989. An extended error model for comparison calibration. *Metrologia*, **26**, 107–113.
David, H. A. 1970. *Order Statistics*. Wiley.
Dieck, R. H. 2007. *Measurement Uncertainty: Methods and Applications*, 4th edn. The Instrumentation, Systems and Automation Society.
Dietrich, C. F. 1973. *Uncertainty, Calibration and Probability*. Adam Hilger.
Dunn, O. J. 1968. A note on confidence bands for a regression line over a finite range. *Journal of the American Statistical Association*, **63**, 1028–1033.

Dunn, P. F. 2005. *Measurement and Data Analysis for Engineering and Science*. McGraw-Hill.

Edwards, A. W. F. 1976. Fiducial probability. *The Statistician*, **25**, 15–35.

Edwards, A. W. F. 1983. Fiducial distributions. Pages 70–76 of: Kotz, S., Johnson, N. L., and Read, C. B. (eds.), *Encyclopedia of Statistical Sciences*, vol. 3. Wiley.

Eisenhart, C. 1963. Realistic evaluation of the precision and accuracy of instrument calibration systems. *Journal of Research of the National Bureau of Standards – C. Engineering and Instrumentation*, **67C**, 161–187.

Elster, C. 2000. Evaluation of measurement uncertainty in the presence of combined random and analogue-to-digital conversion errors. *Measurement Science and Technology*, **11**, 1359–1363.

Elster, C., and Toman, B. 2010. Analysis of key comparisons: estimating laboratories' biases by a fixed effects model using Bayesian model averaging. *Metrologia*, **47**, 113–119.

Elster, C., Chunovkina, A. G., and Wöger, W. 2010. Linking of a RMO key comparison to a related CIPM key comparison using the degrees of equivalence of the linking laboratories. *Metrologia*, **47**, 96–102.

Feldman, G. J., and Cousins, R. D. 1998. Unified approach to the classical statistical analysis of small signals. *Physical Review D*, **57**, 3873–3889.

Feller, W. 1968. *An Introduction to Probability Theory and Its Applications*, 3rd edn, vol. 1. Wiley.

Fisher, R. A. 1930. Inverse probability. *Proceedings of the Cambridge Philosophical Society*, **26**, 528–535.

Fishman, G. S. 1996. *Monte Carlo: Concepts, Algorithms, and Applications*. Springer.

Fisz, M. 1963. *Probability Theory and Mathematical Statistics*, 3rd edn. Wiley.

Fotowicz, P. 2006. An analytical method for calculating a coverage interval. *Metrologia*, **43**, 42–45.

Fraser, D. A. S., Reid, N., and Wong, A. C. M. 2004. Inference for bounded parameters. *Physical Review D*, **69**, 033002.

Frenkel, R. B. 2009. Fiducial inference applied to uncertainty estimation when identical readings are obtained under low instrument resolution. *Metrologia*, **46**, 661–667.

Fuller, W. A. 1987. *Measurement Error Models*. Wiley.

Geisser, S., and Cornfield, J. 1963. Posterior distributions for multivariate normal parameters. *Journal of the Royal Statistical Society, Series B*, **25**, 368–376.

Gleser, L. J. 2002. Comment. *Statistical Science*, **17**, 161–163.

Grabe, M. 2005. *Measurement Uncertainties in Science and Technology*. Springer.

Graybill, F. A., and Bowden, D. C. 1967. Linear segment confidence bands for simple linear models. *Journal of the American Statistical Association*, **62**, 403–408.

Guthrie, W. F., Liu, H., Rukhin, A. L., Toman, B., Wang, J. C. M., and Zhang, N. 2009. Three statistical paradigms for the assessment and interpretation of measurement uncertainty. Pages 71–115 of: Pavese, F., and Forbes, A. B. (eds.), *Data Modeling for Metrology and Testing in Measurement Science*. Birkhäuser.

Hall, B. D. 2006. Computing uncertainty with *uncertain numbers*. *Metrologia*, **43**, L56–L61.

Hall, B. D. 2008. Evaluating methods of calculating measurement uncertainty. *Metrologia*, **45**, L5–L8.

Hannig, J., Iyer, H. K., and Wang, C. M. 2007. Fiducial approach to uncertainty assessment accounting for error due to instrument resolution. *Metrologia*, **44**, 476–483.

Hodges, Jr., J. L., and Lehmann, E. L. 1964. *Basic Concepts of Probability and Statistics*. Holden-Day.

Hoeffding, W. 1982. Asymptotic normality. Pages 139–147 of: Kotz, S., Johnson, N. L., and Read, C. B. (eds), *Encyclopedia of Statistical Sciences*, vol. 1. Wiley.

Howe, W. G. 1969. Two-sided tolerance limits for normal populations – some improvements. *Journal of the American Statistical Association*, **64**, 610–620.

Hughes, I. G., and Hase, T. P. A. 2010. *Measurements and their Uncertainties: A Practical Guide to Modern Error Analysis*. Oxford University Press.

IPCC. 2000. *IPCC Good Practice Guidance and Uncertainty Management in National Greenhouse Gas Inventories. Annex 1: Conceptual Basis for Uncertainty Analysis*, http://www.ipcc-nggip.iges.or.jp/public/gp/english/A1_Conceptual.pdf (accessed 27 January 2012).

ISO. 1998. *SS-EN ISO 14253-1 Geometrical Product Specifications (GPS) – Inspection by Measurement of Workpieces and Measuring Equipment - Part 1: Decision Rules for Proving Conformance or Non-conformance with Specifications*.

ISO. 2004. *ISO/IEC 17000: 2004 Conformity Assessment – Vocabulary and General Principles*.

Jaynes, E. T. 1957. Information theory and statistical mechanics. *Physical Review*, **106**, 620–630.

Jaynes, E. T. 1968. Prior probabilities. *IEEE Transactions on Systems Science and Cybernetics*, **SSC-4**, 227–241.

Jaynes, E. T. 1982. On the rationale of maximum-entropy methods. *Proceedings of the IEEE*, **70**, 939–952.

JCGM. 2008a. *JCGM 100:2008 Evaluation of Measurement Data – Guide to the Expression of Uncertainty in Measurement*, http://www.bipm.org/utils/common/documents/jcgm/JCGM_100_2008_E.pdf.

JCGM. 2008b. *JCGM 101:2008 Evaluation of Measurement Data – Supplement 1 to the "Guide to the Expression of Uncertainty in Measurement" – Propagation of Distributions Using a Monte Carlo Method*, http://www.bipm.org/utils/common/documents/jcgm/JCGM_101_2008_E.pdf.

JCGM. 2011. *JCGM 102:2011 Evaluation of Measurement Data – Supplement 2 to the "Guide to the Expression of Uncertainty in Measurement" – Extension to Any Number of Output Quantities*, http://www.bipm.org/utils/common/documents/jcgm/JCGM_102_2011_E.pdf.

JCGM. 2012. Motivation and Scope for the Revision of the GUM. (Letter sent to National Metrology Institutes, January 2012).

Johnson, N. L., Kotz, S., and Kemp, A. W. 1993. *Univariate Discrete Distributions*, 2nd edn. Wiley.

Johnson, N. L., Kotz, S., and Balakrishnan, N. 1994. *Continuous Univariate Distributions* vol. 1, 2nd edn. Wiley.

Johnson, N. L., Kotz, S., and Balakrishnan, N. 1995. *Continuous Univariate Distributions*, vol. 2, 2nd edn. Wiley.

Johnson, R. A., and Wichern, D. W. 1998. *Applied Multivariate Statistical Analysis*. Prentice Hall.

Kaarls, R. 1980. *Report of the BIPM Working Group on the Statement of Uncertainties (1st meeting 21 to 23 October 1980) to the Comité International des Poids et Mesures*.

Kacker, R., Sommer, K.-D., and Kessel, R. 2007. Evolution of modern approaches to express uncertainty in measurement. *Metrologia*, **44**, 513–529.

Kåhre, J. 2002. *The Mathematical Theory of Information*. Kluwer.

Kiefer, J. 1982. Conditional inference. Pages 103–109 of: Kotz, S., Johnson, N. L., and Read, C. B. (eds.), *Encyclopedia of Statistical Sciences*, vol. 2. Wiley.

Kirkup, L., and Frenkel, R. B. 2006. *An Introduction to Uncertainty in Measurement*. Cambridge University Press.

Larson, H. J. 1982. *Introduction to Probability Theory and Statistical Inference*, 3rd edn. Wiley.

Ledermann, W., and Lloyd, E. (eds). 1984. *Handbook of Applicable Mathematics*, vol. 6, Part B, *Statistics*. Wiley.

Lindgren, B. W. 1968. *Statistical Theory*. Macmillan.

Lira, I. 2002. *Evaluating the Measurement Uncertainty: Fundamentals and Practical Guidance*. Institute of Physics.

Magee, B. 1973. *Popper*. Fontana/Collins.

Mandel, J. 1964. *The Statistical Analysis of Experimental Data*. Dover.

Mandelkern, M. 2002. Setting confidence intervals for bounded parameters. *Statistical Science*, **17**, 149–159.

Marriott, F. H. C. 1990. *A Dictionary of Statistical Terms*, 5th edn. Longman.

Miller, Jr., R. G. 1980. *Simultaneous Statistical Inference*, 2nd edn. Springer-Verlag.

Miller, Jr., R. G. 1986. *Beyond ANOVA, Basics of Applied Statistics*. Wiley.

Murphy, R. B. 1961. On the meaning of precision and accuracy. *Materials Research and Standards*, **4**, 264–267.

Neyman, J. 1937. Outline of a theory of statistical estimation based on the classical theory of probability. *Philosophical Transactions of the Royal Society of London. Series A, Mathematical and Physical Sciences*, **236**, 333–380.

NIST. 2010. http://physics.nist.gov/cuu/Constants/index.html.

NIST. 2012. *NIST/SEMATECH e-Handbook of Statistical Methods*, http://www.itl.nist.gov/div898/handbook/, 2 March 2012.

O'Hagan, A. 1994. *Kendall's Advanced Theory of Statistics*, vol. 2B, *Bayesian Inference*. Edward Arnold.

Pedersen, J. G. 1978. Fiducial inference. *International Statistical Review*, **46**, 147–170.

Popper, K. R. 1968. *The Logic of Scientific Discovery*, 2nd edn. Hutchinson & Co.

Rabinovich, S. 1995. *Measurement Errors: Theory and Practice*. American Institute of Physics.

Rabinovich, S. 2007. Towards a new edition of the "Guide to the expression of uncertainty in measurement". *Accreditation and Quality Assurance*, **12**, 603–608.

Shannon, C. E. 1948. A mathematical theory of communication. *Bell System Technical Journal*, **27**, 379–423, 623–656.

Shirono, K., Tanaka, H., and Ehara, K. 2010. Bayesian statistics for determination of the reference value and degree of equivalence of inconsistent comparison data. *Metrologia*, **47**, 444–452.

Sivia, D. S. 1996. *Data Analysis: A Bayesian Tutorial*. Clarendon Press.

Stevens, S. S. 1946. On the theory of scales of measurement. *Science*, **103**, 677–680.

Stuart, A., and Ord, J. K. 1987. *Kendall's Advanced Theory of Statistics*, 5th edn, vol. 1. Griffin.

van Dyk, D. A. 2002. Comment. *Statistical Science*, **17**, 164–168.

VIM. 2012. *JCGM 200:2012 International Vocabulary of Metrology – Basic and General Concepts and Associated Terms (VIM)*, http://www.bipm.org/utils/common/documents/jcgm/JCGM_200_2012.pdf.

Wang, C. M., and Iyer, H. K. 2005. Propagation of uncertainties in measurements using generalized inference. *Metrologia*, **42**, 145–153.

Wang, C. M., and Iyer, H. K. 2006. Uncertainty analysis for vector measurands using fiducial inference. *Metrologia*, **43**, 486–494.

Wang, C. M., and Iyer, H. K. 2008. Fiducial approach for assessing agreement between two instruments. *Metrologia*, **45**, 415–421.

Wang, C. M., and Iyer, H. K. 2009. Fiducial intervals for the magnitude of a complex-valued quantity. *Metrologia*, **46**, 81–86.

Wang, C. M., and Iyer, H. K. 2010. On interchangeability of two laboratories. *Metrologia*, **47**, 435–443.

Wasserman, L. 2002. Comment. *Statistical Science*, **17**, 163–163.

Wilks, S. S. 1962. *Mathematical Statistics*. Wiley.

Willink, R. 2005. A procedure for the evaluation of measurement uncertainty based on moments. *Metrologia*, **42**, 329–343.

Willink, R. 2006a. An approach to uncertainty analysis emphasizing a natural expectation of a client. Pages 344–349 of: Ciarlini, P. *et al.* (eds.), *Advanced Mathematical and Computational Tools in Metrology* VII. World Scientific.

Willink, R. 2006b. Uncertainty analysis by moments for asymmetric variables. *Metrologia*, **43**, 522–530.

Willink, R. 2007a. A generalization of the Welch–Satterthwaite formula for use with correlated uncertainty components. *Metrologia*, **44**, 340–349.

Willink, R. 2007b. On the uncertainty of the mean of digitized measurements. *Metrologia*, **44**, 73–81.

Willink, R. 2007c. Uncertainty of functionals of calibration curves. *Metrologia*, **44**, 182–186.

Willink, R. 2008. Estimation and uncertainty in fitting straight lines to data: different techniques. *Metrologia*, **45**, 290–298.

Willink, R. 2009. A formulation of the law of propagation of uncertainty to facilitate the treatment of shared influences. *Metrologia*, **46**, 145–153.

Willink, R. 2010a. Difficulties arising from the representation of the measurand by a probability distribution. *Measurement Science and Technology*, **21**, 015110.

Willink, R. 2010b. Measurement of small quantities: further observations on Bayesian methodology. *Accreditation and Quality Assurance*, **15**, 521–527.

Willink, R. 2010c. Probability, belief and success rate: comments on 'On the meaning of coverage probabilities'. *Metrologia*, **47**, 343–346.

Willink, R. 2012a. Measurement uncertainty and procedures of conformity assessment. Pages 426–433 of: Pavese, F. *et al.* (eds.), *Advanced Mathematical and Computational Tools in Metrology and Testing*, vol. 9. World Scientific.

Willink, R. 2012b. Interlaboratory comparisons and the estimation of biases. *Metrologia* (submitted).

Wolfson, D. 1985a. Lindeberg-Feller theorem. Pages 1–2 of: Kotz, S., Johnson, N. L., and Read, C. B. (eds.), *Encyclopedia of Statistical Sciences*, vol. 5. Wiley.

Wolfson, D. 1985b. Lindeberg-Lévy theorem. Pages 2–3 of: Kotz, S., Johnson, N. L., and Read, C. B. (eds.), *Encyclopedia of Statistical Sciences*, vol. 5. Wiley.

Woodroofe, M., and Zhang, T. 2002. Comment. *Statistical Science*, **17**, 168–171.

Index

acceptance sampling vs conformity assessment, 166
accuracy and exactness, statistical, 18, 48, 96, 112
accuracy, in measurement, 7, 9, 67, 204, 214
'all models are wrong', 196
American Institute of Aeronautics and Astronautics, 242
American Society of Mechanical Engineers, 242
ancillary variate, 78
assurance, 3, 20, 45, 86, 249, 253, 257
    postmeasurement, 51, 198, 249, 250, 257
    premeasurement, 51, 198, 249, 257
auxiliary variate, 78

Bayes' theorem, 207–209, 213, 235, 236
Bayesian statistics, 5, 22, 40, 51, 90, 117, 206–214, 239, 240, 243, 244
    anticipated success rate in, 86, 210, 234
    coherence of, 212, 222–227
    controversial nature of, xii, 209
    frequency-based, 213
    meaning of output of, 257
    objective, 96, 212, 213, 220–236, 250–252
    simultaneous inference in, 184
    subjective, 210–212, 214, 228
    view of probability, 32, 86, 209–210
bias, 10, 21, 60, 111, 131, 142, 192, 198, 255
    adjustment for, 126, 129, 135, 201
    as a term, xvii
    definition of, xvii
    estimation of, 198–200
    experimental, 192, 201–202
Bonferroni inequality, 141, 182–184, 195
Bureau International des Poids et Mesures, 57, 237

calibration, 22, 24, 29, 30, 59, 61, 161, 189–191
    of measurement technique, 201
categorical variable, as basis of entropy, 231, 232
central limit theorem, 9, 103
    for identical variables, 52
    for non-identical variables, 24, 53, 54, 65, 67, 104
cherry picking, 144, 184, 255

classical statistics, *see* frequentist statistics
    extended, xv, 73, 81, 258
coherence, *see* Bayesian statistics
combination of data, 254–256
compound distribution, 114–117
confidence, 32, 38, 45–52, 86, 216, 257, 258
    as a term, 45
    average, 25, 29, 55, 72, 73, 80
    conditional, 77–80
    level of, 22, 49, 55, 88, 118, 182–184, 216
confidence band, 187, 188
confidence belt, 247, 248, 250
confidence coefficient, *see* confidence, level of
confidence interval, 32, 40, 45–50, 214
    approximate, definition of, 49
    as a procedure, 46, 93, 112, 198, 247
    as a term, 31, 46
    as random entity, 46, 215
    average, 72, 87, 103, 125, 135, 145–150, 213
        definition of, 74
    conditional, 72, 77–80, 87, 103, 125, 150, 200
        definition of, 78
    definition of, 48, 86
    exact, definition of, 48
    found from hypothesis tests, 184, 197, 247
    left-infinite, 87
    logical basis of, 5, 85
    practical implication of, 46, 62
    preferred to credible interval, 51, 236
    realized, definition of, 46
    right-infinite, 87
    simultaneous, 179–184
    type relevant to status of unknowns, 13
    valid, definition of, 48
    vs probability interval, 215
confidence region, 175–183, 187–189, 229, 266
conformity assessment vs acceptance sampling, 166
conservatism, statistical, 48, 67, 70, 97, 149, 169, 180, 184, 263
consistency, test of, 193–195, 200, 203

consistent gambler, 33, 34, 36, 220, 223, 246
constrained target value, 19, 51, 198, 230, 245–258
correlation, 158, 162, 164
correlation coefficient, 101
covariance matrix, 104, 154–164, 176, 177, 229
credible interval, 5, 51, 210, 214, 235, 236, 251
    definition of, 208
credible, as a term, 45
credible region, 229
cumulants, 105–107
cumulants method, 117–121, 135, 140, 143, 264

data snooping, 184
decision theory, 52
degree of belief, 32, 209–212, 239, 241, 260
    definition of, 34, 38, 220
    frequentist concept of, 239, 244
    long-run requirement for, 35, 86
    vs probability, 38
degrees of freedom, effective number of, 49, 93, 97, 118
digitization, *see* discretization
discretization, 25, 29, 74–76, 224
distributed-measurand concept, 207, 238–241
distribution
    arc-sine, 106, 107, 122, 263
    beta, 67, 107, 118, 264
    bivariate normal, 177
    chi-square, 95, 107, 216
    exponential, 53, 70, 205
    F, 107, 177, 188
    Hotelling's T^2, 176–180, 229
    Laplace, 53, 106
    maximal variance, 67–69
    multivariate normal, 176, 180, 181, 229
    non-central chi-square, 136, 144
    normal, 9, 24, 42, 43, 52, 67, 92, 95, 106, 118, 123, 140, 233, 261, 263, 264
        standard, 52, 96, 107, 108, 122, 168, 247, 261
        truncated, 69, 70, 224–226, 251
    Pearson, 107–111, 116, 118, 121, 264
    raised cosine, 67, 68
    Student's t, 48, 53, 92, 94–97, 105–107, 116, 123, 263
        truncated, 117, 227
    triangular, 66, 106, 113, 115
    uniform, 42, 53, 62, 64, 66, 68, 75, 93, 105–108, 110, 115, 122, 138, 181, 222, 224, 261–265

efficiency, statistical, 50, 94, 140, 141, 175, 178, 184
entropy, 67, 230–233, 240
    as information rate, 231
    meaning with categorical variables, 232
    meaning with numerical variables, 232
environmental quantity, 13, 26, 27, 39, 41, 61, 78–80, 87, 103, 150
error
    calibration, 24, 29, 69
    classification of, 28–30, 237
    definition of, 7
    discretization, 25, 29, 74
    fixed, definition of, 29
    inhomogeneity, 24
    linearization, 27, 112
    'linearized', 105, 117, 118, 120, 123
    matching, 25, 28, 132, 145
    moving, 29, 30, 57, 60, 170, 171
    multiplicative, 229
    numerical vs 'random', 31
    propagation of, 57, 101, 152, 154, 157, 237
    pure, 24, 26, 27
    random, definition of, 28
    sampling, 21, 24, 30
    specification, 27
    statistical, 21, 23, 26, 28, 30
    systematic, definition of, 28
    time-scale for, 169–171
    Type A, 30, 93, 97, 116, 117, 124, 262, 263
    Type B, 30, 93, 97, 114, 117, 262, 263
        estimation of total, 192, 198–200, 203
    vs uncertainty, as a term, 257
error analysis, the subject of, 9, 16, 56, 86
estimate equation
    definition of, 14
    linear approximation to, *see* linear approximation
estimate, as a term, xvii
estimation vs prediction, 6
estimator
    conditional/unconditonal, 199
    consistent, 96
    interval, 47, 103
    linear, 199
    point, 79
    vs estimate, 40
estimator equation, example of, 140
excess, coefficient of, 105–119, 265
    definition of, 105
    interval length related to sign of, 110, 114
    of compound distribution, 115
    of familiar distributions, 105, 106
    of symmetric Pearson distributions, 108
    of t distributions, 116
    propagation of, 107
expected value, 34, 42, 86, 209, 239, 259, 260
experiment equation, definition of, 12
experiment vs measurement, 57, 59–62
experimental uncertainty, 249, 255–257
    as measurement uncertainty, 256

fiducial inference, 205–207, 239, 243
Fisher, R. A., Sir, 52, 205, 206, 235
fixed estimate, 22, 25, 39, 77, 102, 130–135, 145–147, 150
fluctuating quantity, 6, 12, 26, 78, 244
frequentist statistics
    as 'classical' statistics, 21

concept of 'degree of belief' in, 210
  extended, xv, 73, 81, 258
  in revision of the *Guide*, 243
  practical, 47, 73, 76
  textbook, 73, 76, 204, 206, 249
  view of probability, 32, 35
function estimation, 6, 161, 185–191

Gauss, C. F., 235
goodness-of-fit test, 192, 196–197
Guide to the Expression of Uncertainty in Measurement, xii, 5, 13, 18, 29, 58, 93, 97, 118, 148, 237–244

hypothesis, simple, 196
hypothesis test, 51, 167, 182, 184, 193, 196, 197, 200
  as detection, not measurement, 253
  inversion to obtain a confidence interval, 184, 197, 247

ignorance, intention to quantify, 222, 227–230
indication, as a term, 244
inference, statistical, definition of, 45
information, 17, 28, 50, 51, 208, 220, 230–233
  as a term, 221
  as an objective concept, 212, 220–221
  contained in ordering and distance, 232
  rate of gain of, 231
  vs information rate, 231
input quantity, as a term, 241, 244
Intergovernmental Panel on Climate Change, 242
International Vocabulary of Metrology, xvii, 5, 17, 214
interval
  confidence, definition of, 48, 86
  credible, definition of, 208
  prediction, example of, 189
  probability, definition of, 215
  tolerance, definition of, 216

kurtosis, 105, 123

Laplace, P.-S., 235
least squares, 187
likelihood function, 208, 211
linear approximation, 27, 98–125, 128–132, 136, 144, 157

mathematical expectation, *see* expected value
measurand
  definition of, xv, 5
  identification of, 10–12
measurand equation, 12–16
  as an approximation, 14, 27
  contribution to error, 27
  decomposition of, 15, 130
  definition of, 12

implicit, 15, 111
  linear approximation to, *see* linear approximation
measurement comparison, 192–203, 254
measurement error
  as secondary concept for a statistician, 21
  definition of, 7, 85
measurement estimate, definition of, 7
measurement result, definition of, 7
measurement vs experiment, 57, 59–62
measurement, goal of, 4, 20
model
  approximate nature of, xiv, 14, 27, 49, 196
  inadequacy of, 254
  of measurement, 16, 88–90
  purpose of, 185
  statistical/probabilistic, 14, 38, 43–45
  subject of, 38
Monte Carlo analysis, 55, 88, 97, 98, 151, 181
Monte Carlo estimation of quantile, 120
Monte Carlo evaluation of distribution, 53, 110, 120–124, 182, 184
Monte Carlo integration, 54
Monte Carlo simulation, *see* simulation of the measurement
multiple comparisons, 182

Neyman, J., 51, 52, 205
non-linear function, 27, 125–151
normal-approximation method, 105, 114, 116–119, 135, 140, 143
notation, xiv, 14, 39–41
  non-standard, 44, 106
  postmeasurement, 43, 85, 89, 90, 103, 158
  premeasurement, 43, 85, 89, 90, 194
  statistical vs non-statistical, 41

observation, as a term, 244

parameter
  definition of, 13
  estimation of, xii, xiv, xvii, 6, 12, 47, 87
  of the experiment, 13
parent, as a term, 44, 90
Pearson system of distributions, 107–109
Pearson, E., 51, 52
Pearson, K., 107
Popper, K. Sir, 37
posterior distribution, 208, 222, 234, 252
  marginal, 230
  predictive, 225
prediction interval, 189
prediction vs estimation, 6
predictive distribution, 224, 225
principle of maximum entropy, 67, 231–233, 240
prior distribution, 76, 205, 208, 209, 229, 235
  as an inconvenience, 236
  elicitation of, 211
  formed from a frequency distribution, 213

    improper, 96, 222–227, 231, 250
    mandatory in Bayesian analyses, 207, 239
    matching, 229, 230, 235
    reference, 233
    unable to correctly represent ignorance, 228
    with delta function, 252
probability
    actual vs nominal figure, 50
    as a term, 219, 243
    as degree of belief, 34, 209
    as long-run concept, 34, 56
    as requiring an event or hypothesis, 5
    Bayesian concept of, 32, 207
    conditional, 78, 79
    controversy about, xii
    definition of, 34–36, 38
    fiducial, 205
    for prediction instead of estimation, 6
    frequentist and Bayesian scopes of, 36–38, 209–210
    frequentist concept of, 32, 35, 38
    in classical statistics, 21
    lack of definition of, 234
    legitimate subjects of, 32, 36–38
    long-run implication of, 34, 86
    personal/subjective, 34, 210, 260
    posterior, 209, 224, 225, 234, 252
    potential presupposition about, 235–236
    practical implication of, 46, 47
    practical meaning required for, 32
    prior, 51, 52, 213, 222, 223
    requires a long-run context, 124
    statement, example of, 5
    vs belief, 38
probability density function, notation, 42
probability interval, 215
product testing, 164–169, 253

quantile, definition of, 45
quantity, as a term, xvii, 241, 244

random, as a term, 31, 42
random variable
    as a term, 42
    simple definition of, 39
    variate as synonym for, 42
rational agent, *see* consistent gambler
reasonableness distribution, 217–218
reference analysis, 233
reference values, 25, 29, 44, 61, 132, 148, 215
regression, 21, 22, 186–189,
regression function, 186
rejection region, 247–250
repetition
    broadening of concept of, 22, 31, 60, 201
    of the experiment, 60, 201
    of the measurement, 60, 201

sample, as a term, 24
significance level, 167, 195, 203
    adjustment for simultaneous inference, 195
significance test, 167
simulation of the measurement, 13, 50, 53–55, 74–76, 94, 138–151, 181–183, 262, 263
simultaneous inference, 175, 179–184, 187, 195
Sleeping Beauty paradox, 64, 260
source-coding theorem, 231
standardized distribution, 107
success rate, 17, 35, 47, 60, 170, 210, 240, 243, 257
    perception of, 63–65, 71, 76, 86
    practical, 62, 73, 77, 112, 234
success, definition of, xiv, 17, 19, 86, 175

target value
    definition of, 6, 19
    measurement without concept of, 6, 214–219
    uncertainty without concept of, 10, 214–219
Taylor's series, 27, 98, 125, 135
time-scale, 28–30, 169–171
tolerance interval, 216–217
transfer of uncertainty information, 14, 150
true value, xii, 4–7, 17, 37, 214
    uncertainty without concept of, 10
Type I error probability, partitioning of, 148

uncertainty (of measurement)
    as 'experimental uncertainty', 256
    as a term, 57, 214, 219, 237, 255–257
    as being specific to observer, 65
    as potential magnitude of error, 8, 86
    dependent on purpose, 152–171, 187, 255
    information about, 14, 118, 150, 152–171
    interval, meaning of, 16–19, 35, 86, 244
    language about, 10
    non-uniqueness of, 15, 124, 256
    problematic nature of, xii
    propagation of, 57, 101, 130, 157, 237
    standard, 8–12, 58, 130, 169–170, 215, 243, 257
    Type A evaluation of, 29, 124, 239, 240, 243
    Type B evaluation of, 29, 93, 239–241, 243–244
    without concept of target value, 10, 214–219

validity, statistical, 15, 48, 96, 140, 149, 175, 184
value, as a term, xvii, 219, 244
variate
    as synonym for random variable, 42
    categorical, 231, 232
    continuous numerical, 42, 232, 233
    discrete numerical, 232

weak law of large numbers, 35, 259
Welch–Satterthwaite formula, 93
Working Group (on the statement of uncertainties), 57–59, 62, 65, 70, 86, 166, 237, 240, 242, 243
worst-case values, xii, 56, 60, 69, 70, 72, 166, 169, 204